THE WISDOM OF THE BODY

WALTER BRADFORD CANNON was George Higginson Professor of Physiology at Harvard Medical School from 1906 to 1942, leaving briefly on several occasions, once to serve with the British and American Expeditionary Forces in 1917–18 and later as Exchange Professor to France in 1929 and as Visiting Professor to Peiping in 1935. His researches dealt with the effects of emotional excitement, internal secretions and organic stabilization, and were first formulated in his book *Bodily Changes in Pain, Hunger, Fear and Rage,* a classic of modern physiology. Out of his war experiences came his famous study, *Traumatic Shock,* and research on stomach and intestinal movements led to *Digestion and Health.*

Dr. Cannon was widely honored. He received honorary doctorates from Yale University, Boston University, and the Universities of Liege, Strasbourg and Paris, as well as from Harvard. He was a member of the American Academy of Arts and Sciences and of the National Academy of Sciences, and was President of the American Association for the Advancement of Science.

The WISDOM of THE BODY

WALTER B. CANNON M.D., SC.D.

The Norton Library
W · W · NORTON & COMPANY · INC ·
NEW YORK

FIRST PUBLISHED IN THE NORTON LIBRARY 1963

Books That Live
The Norton imprint on a book means that in the publisher's estimation it is a book not for a single season but for the years.
W. W. Norton & Company, Inc.

ISBN 0 393 00205 5
PRINTED IN THE UNITED STATES OF AMERICA
6 7 8 9 0

TO

RALPH BARTON PERRY

GUIDE PHILOSOPHER FRIEND

CONTENTS

LIST OF ILLUSTRATIONS

PREFACE

THE course of the researches with which I have been concerned during more than a third of a century devoted to physiological investigation seems to have been consistent and continuous. Almost the first research, which was undertaken when I was a student of medicine, was a study of the phenomena of swallowing. Next I was led quite naturally to observe the motions of the stomach, then the intestines, and the conditions affecting their activities. The volume, "The Mechanical Factors of Digestion," which summarized about ten years of work on the digestive canal, closed with chapters on the nervous control of the digestive processes and the ways in which emotional states may be disturbing. The succeeding group of investigations, concerned with the influence of emotional excitement on adrenal secretion and with the significance of the attendant changes in the organism, originated in the earlier observations on the influence of emotions on digestion. Those investigations were reported as a whole in the volume, "Bodily Changes in Pain, Hunger, Fear, and Rage." They led in turn to an interest in the general functions of the autonomic nervous system—an interest which was fostered by researches carried on during the War that were described in the book, "Traumatic Shock."

In the present volume another step in the natural sequence of ideas is apparent. In the main it pertains to the relation of the autonomic system to the self-regulation of physiological processes. That relation was only slowly disclosed. Indeed, not a few researches on the service of the autonomic in providing for stability of the organism had been completed and published before the connection of that system with regulatory arrangements was clearly understood. We found that we had long been working on the rôle of the autonomic system in maintaining steady states without realizing that we were doing so! Then facts already discovered took on new significance. Thus considerable work of an early date became pertinent, and it has been included in the pages which follow.

The main features of this book were first presented in a technical article in "Physiological Reviews" in 1929, under the title, "Organization for Physiological Homeostasis." The relation of the autonomic system to the regulation of steady states in the body was outlined in the Linacre Lecture for 1930, at the University of Cambridge. The ideas in these two presentations were amplified in a series of lectures at the Sorbonne in the winter of 1930. They are offered here for the general reader because of their interest to others than biologists. I hope, however, that this exposition may prove suggestive to biologists and investigators as well, for I have taken occasion to point out many gaps in our knowledge where further research may be useful.

In 1923, the late Professor E. H. Starling, of University College, London, gave the Harveian Oration before the Royal College of Physicians. He paid high tribute to William Harvey for emphasizing and demonstrating the great value of the experimental method for the solution of biological problems. And he expressed eloquently his admiration for the marvelous and beautiful adjustments in the organism, that had been revealed by following Harvey's injunction to "study out the secrets of nature by way of experiment." His oration he entitled "The Wisdom of the Body." Only by understanding the wisdom of the body, he declared, shall we attain that "mastery of disease and pain which will enable us to relieve the burden of mankind." Because my own convictions coincide with those of Professor Starling, and because the facts and interpretations which I shall offer illustrate his point of view, I have chosen to give the title of his oration to the present volume.

WALTER B. CANNON

Boston, 1932

PREFACE TO THE SECOND EDITION

SEVEN years have passed since the first edition of this book was published. During that time additional researches concerned with homeostasis have been completed; some criticisms have been received and answered; and extensions of the concept have been undertaken.

In 1938 there came a request for a translation of the book into French. Naturally an inclusion of the newer developments in the realm of homeostasis was desirable. Among these were studies on the dog, and the revelation of the remarkable powers of preserving stability manifested by that animal when he is exposed to unfavorable conditions. Also an inquiry into the efficacy of homeostatic mechanisms at different age periods showed gradual limitations as the years of life advance toward its end; these facts were evidently pertinent to understanding the nature of the aging process. Furthermore, the advances of physiology naturally produced fresh information which could be added to that which was included in the first edition.

The prospect of a French edition of "The Wisdom of the Body," which should include the latest results of studies on the regulation of stability in the fluid matrix, led to a consultation with Mr. Norton regarding the possibility of a simultaneous issue of the French translation and a second English edition. Very graciously he agreed to the plan.

In this second edition, therefore, new facts and developments in the realm of bodily stabilization have been incorporated, in order to bring the exposition up to date; and also a new chapter has been added, on the effects of age on homeostatic mechanisms.

WALTER B. CANNON

Harvard Medical School
Boston, Mass., 1939

INTRODUCTION

I

OUR bodies are made of extraordinarily unstable material. Pulses of energy, so minute that very delicate methods are required to measure them, course along our nerves. On reaching muscles they find there a substance so delicately sensitive to slight disturbance that, like an explosive touched off by a fuse, it may discharge in a powerful movement. Our sense organs are responsive to almost incredibly minute stimulations. Only recently have men been able to make apparatus which could even approach the sensitiveness of our organs of hearing. The sensory surface in the nose is affected by vanillin, 1 part by weight in 10,000,000 parts of air, and by mercaptan 1/23,000,000 of a milligram in a liter (approximately a quart) of air. And as for sight, there is evidence that the eye is sensitive to 5/1,000,000,000,000 erg, an amount of energy, according to Bayliss, which is 1/3,000 that required to affect the most rapid photograph plate.

The instability of bodily structure is shown also by its quick change when conditions are altered. For example, we are all aware of the sudden stoppage of action in parts of the brain, accompanied by fainting and loss of consciousness, that occurs when there is a momentary check in the blood flow through its vessels. We know that if the blood supply to the brain wholly ceases for so short a time as seven or eight minutes certain cells which are necessary for intelligent action are so seriously damaged that they do not recover. Indeed, the high degree of instability of the matter of which we are composed explains why drowning, gas poisoning, or electric shock promptly brings on death. Examination of the body after such an accident may reveal no perceptible injury that would adequately explain the total disappearance of all the usual activities. Pathetic hope may rise that this apparently normal and natural form could be stirred to life again. But there are subtle changes in the readily mutable stuff of the human organism which prevent, in these conditions, any return of vital processes.

When we consider the extreme instability of our bodily structure, its readiness for disturbance by the slightest application of external forces and the rapid onset of its decomposition as soon as favoring circumstances are withdrawn, its persistence through many decades seems almost miraculous. The wonder increases when we realize that the system is open, engaging in free exchange with the outer world, and that the structure itself is not permanent but is being continuously broken down by the wear and tear of action, and as continuously built up again by processes of repair.

II

The ability of living beings to maintain their own constancy has long impressed biologists. The idea that disease is cured by natural powers, by a *vis medicatrix naturae*, an idea which was held by Hippocrates (460–377 B. C.), implies the existence of agencies which are ready to operate correctively when the normal state of the organism is upset. More precise references to self-regulatory arrangements are found in the writings of modern physiologists. Thus the German physiologist, Pflüger, recognized the natural adjustments which lead toward the maintenance of a steady state of organisms when (1877) he laid down the dictum, "The cause of every need of a living being is also the cause of the satisfaction of the need." Similarly, the Belgian physiologist, Léon Fredericq, in 1885, declared, "The living being is an agency of such sort that each disturbing influence induces by itself the calling forth of compensatory activity to neutralize or repair the disturbance. The higher in the scale of living beings, the more numerous, the more perfect and the more complicated do these regulatory agencies become. They tend to free the organism completely from the unfavorable influences and changes occurring in the environment." Again, in 1900, the French physiologist, Charles Richet, emphasized the remarkable fact. "The living being is stable," he wrote. "It must be so in order not to be destroyed, dissolved or disintegrated by the colossal forces, often adverse, which surround it. By an apparent contradiction it maintains its stability only if it is excitable and capable of modifying itself according to external stimuli and adjusting its response to the stimulation. In a sense it is stable because it is modifiable—the slight instability is the necessary condition for the true stability of the organism."

Here, then, is a striking phenomenon. Organisms, composed of material which is characterized by the utmost inconstancy and unsteadiness, have somehow learned the methods of maintaining constancy and keeping steady in the presence of conditions which might reasonably be expected to prove profoundly disturbing. For a short time men may be exposed to dry heat at 115 to 128 degrees Centigrade (239 to 261 degrees Fahrenheit) without an increase of their body temperature above normal. On the other hand arctic mammals, when exposed to cold as low as 35 degrees Centigrade below freezing (31 degrees below zero Fahrenheit) do not manifest any noteworthy fall of body temperature. Furthermore, in regions where the air is extremely dry the inhabitants have little difficulty in retaining their body fluids. And in these days of high ventures in mountain climbing and in airplanes human beings may be surrounded by a greatly reduced pressure of oxygen in the air without showing serious effects of oxygen want.

Resistance to changes which might be induced by external circumstances is not the only evidence of adaptive stabilizing arrangements. There is also resistance to disturbances from within. For example, the heat produced in maximal muscular effort, continued for twenty minutes, would be so great that, if it were not promptly dissipated, it would cause some of the albuminous substances of the body to become stiff, like a hard-boiled egg. Again, continuous and extreme muscular exertion is accompanied by the production of so much lactic acid (the acid of sour milk) in the working muscles that within a short period it would neutralize all the alkali contained in the blood, if other agencies did not appear and prevent that disaster. In short, well-equipped organisms—for instance, mammalian forms—may be confronted by dangerous conditions in the outer world and by equally dangerous possibilities within the body, and yet they continue to live and carry on their functions with relatively little disturbance.

III

The statement was made above that somehow the unstable stuff of which we are composed had learned the trick of maintaining stability. As we shall see, the use of the word "learned" is not unwarranted. The perfection of the process of holding a stable state in spite of extensive shifts of outer circumstance is not a special gift bestowed upon the highest organisms but is the consequence of a gradual evolution. In the eons of time during which animals have developed on the earth probably many ways of protecting against the forces of the environment have been tried. Organisms have had large and varied experience in testing different devices for preserving stability in the face of agencies which are potent to upset and destroy it. As the construction of these organisms has become more and more complex and more and more sensitively poised, the need for more efficient stabilizing arrangements has become more imperative. Lower animals, which have not yet achieved the degree of control of stabilization seen in the more highly evolved forms, are limited in their activities and handicapped in the struggle for existence. Thus the frog, as a representative amphibian, has not acquired the means of preventing free evaporation of water from his body, nor has he an effective regulation of his temperature. In consequence he soon dries up if he leaves his home pool, and when cold weather comes he must sink to its muddy bottom and spend the winter in sluggish numbness. The reptiles, slightly more highly evolved, have developed protection against rapid loss of water and are therefore not confined in their movements to the neighborhood of pools and streams; indeed, they may be found as inhabitants of arid deserts. But they, like the amphibians, are "cold-blooded" animals, that is, they have approximately the temperature of their surroundings, and therefore during the winter months they must surrender their active existence. Only among the higher vertebrates, the birds and mammals, has there been acquired that freedom from the limitations imposed by cold that permits activity even though the rigors of winter may be severe.

The constant conditions which are maintained in the body might be termed *equilibria*. That word, however, has come to have fairly exact meaning as applied to relatively simple physico-chemical states, in closed systems, where known forces are balanced. The coördinated physiological processes which maintain most of the steady states in the organism are so complex and so peculiar to living beings—involving, as they may, the brain and nerves, the heart, lungs, kidneys and spleen, all working coöperatively—that I have suggested a special designation for these states, *homeostasis*. The word does not imply something set and immobile, a stagnation. It means a condition—a condition which may vary, but which is relatively constant.

It seems not impossible that the means employed by the more highly evolved animals for preserving uniform and stable their internal economy (i.e., for preserving homeostasis) may present some general principles for the establishment, regulation and control of steady states, that would be suggestive for other kinds of organization—even social and industrial—which suffer from distressing perturbations. Perhaps a comparative study would show that every complex organization must have more or less effective self-righting adjustments in order to prevent a check on its functions or a rapid disintegration of its parts when it is subjected to stress. And it may be that an examination of the self-righting methods employed in the more complex living beings may offer hints for improving and perfecting the methods which still operate inefficiently and unsatisfactorily. At present these suggestions are necessarily vague and indefinite. They are offered here in order that the reader, as he continues into the concrete and detailed account of the modes of assuring steady states in our bodies, may be aware of the possibly useful nature of the examples which they offer.

IV

In the chapters which follow I propose to consider, first, what may be regarded as the fundamental condition of stability, then the various physiological arrangements which serve to restore the normal state when it has been disturbed, and finally, the narrowing limits of adaptation imposed by age. While considering these arrangements we shall gradually become acquainted with the general devices which are employed to regulate and control the numerous processes and the supplies of material required for our natural activities. We shall see that the nervous system is divisible into two main parts, the one acting outwardly and affecting the world about us, and the other acting inwardly and helping to preserve a constant and steady condition in the organism itself. I shall strive to describe the physiological agencies and events in terms which will be clear to anyone who has had a simple training in biology and in general science.

References

Bayliss. Principles of General Physiology. London, 1915.
Fredericq. Arch. de Zoöl. Exper. et Gén., 1885, iii, p. xxxv.
Pflüger. Pflüger's Arch., 1877, xv, 57.
Richet. Dictionnaire de Physiologie, Paris, iv, 72.

1

THE FLUID MATRIX OF THE BODY

I

WE ORDINARILY speak of ourselves as air-inhabiting animals. A little reflection will disclose, however, the interesting fact that we are separated from the air which surrounds us by a layer of dead or inert material. The skin has an outer covering of dry and horny scales (which may, of course, at times be wet with sweat), and the surfaces of the eyes and the inner parts of the nose and the mouth are bathed in a salty water. All of us that is *alive*, the vast multitudes of minute living elements or cells which compose our muscles, glands, brain, nerves and other parts, reside within this surface coat of non-living stuff. And, except their sides where they are contiguous one with another, the cells are in contact with *fluid*. The living elements of the body, therefore, are water inhabitants, or inhabitants of water which has been modified by the addition of salt and thickened by an albuminous or colloid material. In order to understand the significance of this watery environment or fluid matrix, we must inquire into the services it performs and how it performs them.

To the simple organisms which may be found attached to the rocks of the bed of a stream the flowing water brings the food and oxygen needed for existence and carries away the waste. These single-cell creatures can live only in watery surroundings; if the stream dries they die or enter a dormant state. Similar conditions prevail for the incalculable myriads of cells which constitute our bodies. Each cell has requirements like those of the single cell in the flowing stream. The cells of our bodies, however, are shut away from any chances to obtain directly food, water and oxygen from the distant larger environment, or to discharge into it the waste materials which result from activity. These conveniences for getting supplies and eliminating debris have been provided by the development of moving streams within the body itself—the blood and lymph streams. They work together to carry food, water and oxygen away from the moist surfaces of the body and to deliver these necessities to the cells situated even in the remotest nooks of the organism. From these cells in turn they bring back to the moist surfaces, in the lungs and kidneys, the useless waste of cellular activity which must be discharged.

The movements of the blood and lymph are related to each other somewhat as the movement of water in a rivulet is related to the more stagnant water in the swamp through which it flows. The blood passes along fixed courses in tubular vessels; the lymph or tissue fluid, which fills all the chinks and crannies of the body structure, outside the blood vessels, until it too is gathered in its own channels, is shifted slowly from place to place. We shall now examine the nature of these fluids and the ways in which the internal, proximate environment of the cells is made favorable by keeping the fluids on the move and constantly fresh and uniform.

II

The blood, which constitutes about 8 per cent of our body weight, is a remarkable fluid. It consists of immense numbers of red corpuscles (the normal content of 1 cubic millimeter of a man's blood is 5,000,000 of these corpuscles) and also of many minute motile white corpuscles, all of which float in a thickish watery solution of salts, sugar and albuminous material, the plasma. The red corpuscles play a vital rôle in the body because in the lungs they are able to take on very quickly a practically full load of oxygen and then to unload it more or less completely in other parts of the body where the cells are in need of it. On the way back from these cells to the lungs the red corpuscles help to carry one of the waste products of activity, the carbon dioxide arising from the oxidation which yields heat and which is essential for mechanical work in the functioning of the organism. The motile white corpuscles serve as scavengers and protectors against inert foreign particles and invading germs, which, if they were permitted to accumulate, would pollute the stream.

The plasma, which amounts to more than half of the total mass of the blood, is a conveyor of all manner of food materials provided by the final digestive processes in the intestines. These materials, like oxygen, are carried to all parts of the organism, so that every cell, even the most secluded, shall receive its proper supply, or, if they are not needed, they are brought to special organs in the body where they are stored for future use. Another function of the plasma is to carry from the cells everywhere the waste substances, apart from carbon dioxide, which result from the workings of the bodily machine, and to deliver them to the kidneys through which they are discharged.

The plasma also has the remarkable capacity to change from a fluid to a jelly—to clot or coagulate—when it comes into contact with an injured region. If, for example, the blood vessels are damaged or cut, and there is danger of loss of blood through the opening, the jellifying or clotting of the plasma forms a plug which more or less promptly closes the opening and prevents what might otherwise be a serious hemorrhage.

The lymph differs from blood chiefly in containing no red corpuscles and less albuminous matter than the plasma. It does contain, however, white corpuscles, and also sugar and salts. And it is capable of clotting, though the clot formed by lymph is a less firm jelly than that normally formed by the blood itself.

Since the lymph, or tissue fluid, lies between the blood vessels and the tissue cells, all the substances exchanged between the cells and the flowing blood must pass through it. It is, therefore, the direct intermediator of that exchange.

The distinction between blood and lymph everyone has had occasion to observe in slight injuries of the skin. An accidental blow or pinch may involve only the superficial layers, and then there is produced a "water blister," filled with lymph. If deeper layers of the skin are injured blood vessels are broken and the blood which escapes makes a "blood blister."

In the foregoing and in later sections the term "lymph" covers not only the fluid in lymph vessels but also extracellular, extravascular "tissue fluid." When body "water" is mentioned it includes dissolved salts.

III

Because blood and lymph are limited in amount, the only way in which they can serve continuously as carriers between fixed, secluded cells and the transmitting surfaces of the body is by being used over and over again. They must circulate (see fig. 1). The blood is forced through the vessels by the contractions or "beats" of the heart—essentially a powerful, hollow muscle having two main chambers, right and left. Each of these chambers has tough, membranous inlet and outlet valves. The organization in the heart muscle is such that after each contraction it is required to rest before it is able to beat again. Thus, though the heart is

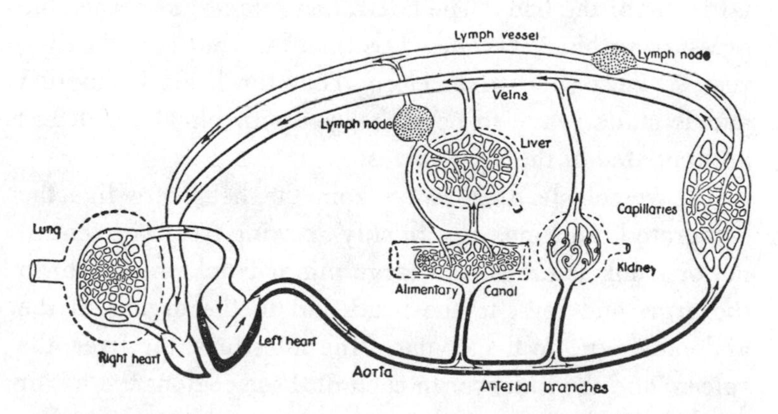

Fig. 1. Diagram of the circulatory and lymphatic systems. The left heart chamber pumps blood out into arteries which distribute it to capillaries. Venous blood is collected from capillaries and returned to the right heart chamber by veins. Thence it is pumped to the lungs and onward to the left heart chamber. Tissue fluid (lymph), exuded through capillary walls, is collected in lymph vessels and returned to veins near the heart (modified from Paton).

continuously at work, beating about 60 times a minute or faster, and with each beat driving forth a heavy load of blood, it may keep at its task for seventy years or more without any obvious fatigue. During the rest period after each contraction blood flows into the chambers of the hollow muscle through the inlet valves—into the right chamber from all the remote regions of the body, and into the left chamber from the lungs. When next the muscle contracts and tightens on its contents, these valves close and prevent a back-flow. The pressure now rises on the contents until it opens the outlet valves, whereupon the blood is driven forth through these valves into the outleading vessels—from the right chamber into the vessels distributing to the lungs, and from the left chamber into the great, main vascular trunk of the body. The heart then relaxes, and when the pressure within it becomes less than that in the outleading vessels, the outlet valves close. Thus the heart is emptied and is made ready to be recharged with blood which has accumulated at the inlet valves.

The vessels leading away from the heart are like the elaborate branchings of a thickly growing tree. The central major trunk is the aorta. Large minor trunks reach out to the arms and legs, to the head and to the organs of the abdomen, e. g., to the stomach, the intestines, the liver, the spleen, and the kidneys. In each of these regions the minor trunks ramify again and again into smaller and smaller branches and twigs until every part of the body is supplied. The vessels leading away from the heart are known as arteries, and this intricate branching system is sometimes called the "arterial tree." The arteries have relatively thick elastic walls, which, because they are provided with an encircling muscular layer, have a variable capacity. When the heart discharges its load of blood into the arterial tree it starts a distending wave along the stream of blood already contained within the system, a wave which can be felt in any superficial branch—in the wrist at the base of the thumb, at the temple in front of the ear, or back of the ankle on the inner side of the foot—as a "pulse."

We must remember always that the virtue of the circulating blood is its service to the cells of the organism which are far removed from the sources of material supplies and from the excretory surfaces where they can get rid of their rubbish. Obviously this service must be performed through the walls of the vessels in which the blood flows. The walls of the arteries are too thick to permit the passage of material to and fro. The process of exchange takes place through the walls of "capillaries," which are extremely minute tubules with walls so exceedingly tenuous that gases, such as oxygen and carbon dioxide, and sugar and salts, in watery solution, readily pass through them. The capillaries, about 1/4000 of an inch in diameter, form a finely reticulate mesh, intimately insinuated between the layers and masses of the cells everywhere in the body. Almost every point pricked with a needle will ooze blood. Into this fine mesh the smallest twigs of the arterial tree, the arterioles, pour the blood; and from it the blood is gathered into the fine twigs of another tree, the tree of veins. From the venules (corresponding to the arterioles) the blood flows to larger and larger veins, with thicker and stronger walls, until the main trunks are reached, the inferior and superior venae cavae, which pour the blood accumulated from all parts of the body into the right chamber of the heart.

In several parts of the body, but notably in the abdomen, the veins may divide into a second set of capillaries which in turn empty into a second set of veins. Thus the blood from the abdominal portion of the alimentary canal and from the pancreas and spleen enters the "portal vein" and flows to the liver; there it enters the hepatic capillaries and only after traversing them does it reach the true hepatic veins and move directly to the heart (see fig. 1).

A system of arteries and veins connects the capillaries of the lungs with the heart (see fig. 1). The essential feature of the service of the circulation in the lungs as in all other parts, it is important to note, is the flow of the blood through the capillaries. Only in the capillary region do the necessary changes occur. All of the rest of the circulatory system exists to maintain the flow in that region, where the blood is useful to the cells.

IV

Lymph is produced by the filtering of a portion of the plasma through the capillary wall. In some parts of the body, for example in the liver, the capillaries are so "permeable" that the process of filtration occurs continuously; in other parts, for example in the limbs, it occurs only when the organs are active. Under such conditions the lymph may be formed more rapidly than it can be carried away, and the part may thus become perceptibly larger.

The lymph is returned to the blood in two quite different ways. The watery portion may pass back to some degree through the capillary wall when the activity of the organ ceases and the filtration pressure in the capillaries is consequently reduced; or the lymph as a whole may enter a definite system of very thin-walled tubes, the so-called "lymphatics," and be conducted through them to a large vein near the heart where the lymph is delivered, as a stream, into the blood (see fig. 1). The larger lymphatic vessels, like the veins, are provided with valves—cup-shaped sacs attached at one side—which prevent a back-flow away from the heart. In consequence every, even slight, pressure exerted on the vessels pushes the contents onward toward the exit. In their course the lymphatic vessels are interrupted by nodes or "glands," which act as sieves and hold back small particles, such as bacteria, which may have found entrance into a tissue space, and keep them from being widely spread through the rest of the body. When protecting the body in this way they become enlarged and can then be felt as swollen, tender lumps.

V

The multitudes of finely branched arterioles, through which the blood must be driven on its way to the capillaries, offer a considerable frictional resistance. When the heart beats and discharges its contents its muscular walls must develop a pressure which will force the blood not only past this resistance but also through the capillary mesh and the veins. With each fresh delivery of blood from the heart the elastic arteries stretch to accommodate the extra contents, and while the heart muscle, behind the barrier set by the outlet valves (see fig. 1), is resting and being filled, the elastic recoil of the stretched arterial walls presses the blood continuously onward. Measurements show that the blood in the arteries is moving under a fairly high pressure, equal, in young adults, to a column of about 120 millimeters of mercury (about 5 feet of water) at the peak of the cardiac discharge into them, i.e., the systolic pressure, and to about 80 millimeters of mercury just before the next discharge, i.e., the diastolic pressure. In the capillaries the pressure has fallen to about 25 millimeters (about 12 inches of water), and it falls progressively as it courses through the veins until its lowest point is found as the blood enters the right chamber of the heart.

Clearly the same amount of blood must pass through the heart, the lungs, the arteries, capillaries and veins during the same period of time, or otherwise the circulation could not continue. Since the total cross-area of the capillaries is much greater than the cross-area of the aorta or the large veins leading to the heart, the blood moves much more slowly in the capillaries than in either the large arterial or venous trunks. This slow flow in the capillary region gives time for the important exchange between the blood and the tissue cells that occurs there.

As we shall soon see, the circulation must vary greatly in its service to the needy cells, according to their degree of activity. The adjustments are brought about chiefly through the nervous control of the heart and blood vessels. The heart can be made to beat more slowly by extra action of the vagus nerves (see fig. 17), which regularly hold the heart rate in continuous or "tonic" check; it can be made to beat more rapidly by action of the sympathetic nerves, and also, interestingly, by lessening the vagal tone. The blood vessels, especially the arterioles, are likewise under control of the sympathetic and other nerves, which cause contraction or relaxation of the muscle in the vascular wall, thus limiting the flow to one part and distributing a larger volume to another part as need arises. Indeed, in special circumstances the mass of the blood can be partially shifted from one region of the body to other regions in an adaptive manner.

We shall encounter numerous instances of the ways in which the sympathetic nervous system acts to modify and adjust conditions in the body so as to preserve constancy and stability. It seems best to consider the general organization of this system at a later stage of our discussion (Chapter XV), when its services can be surveyed as a whole. If the reader is not already acquainted with the main features of the system and at any stage finds the references to it not quite clear, he should turn to Chapter XV and read it.

VI

The facts cited in earlier pages to illustrate the stability of organisms, when exposed to disturbing external and internal conditions, raised questions as to the manner in which that stability is attained. It was the great French physiologist, Claude Bernard, who first suggested that a highly important factor in the establishment and maintenance of steady states in the body is the internal environment, or what we have called the fluid matrix. As early as 1859–60, Bernard pointed out in his lectures that there are two environments for complex living beings—a general environment which is the same as that for inanimate objects and which surrounds the organism as a whole, and an internal environment in which the living elements of the body find their optimal habitat. He first regarded the plasma of the blood as the sole internal environment. Later he spoke of the plasma and the lymph as constituting the *milieu interne*. Finally, in his treatise on the phenomena of life, he referred to it as "the totality of the circulating fluids of the organism."

It was a signal contribution to our understanding of physiology that Bernard made when he recognized that the blood and the interstitial lymph provide appropriate and favorable surroundings for the living cells of the organism. He early pointed out that the *milieu interne* not only is a vehicle for carrying nourishment to cells hidden away in the deep tissues, far from the surfaces of contact with the world outside, and for bringing away from these cells refuse for excretion, but also that it is under control of agencies which keep it remarkably constant. He clearly perceived that just insofar as that constancy is maintained, the organism is free from external vicissitudes. "It is the fixity of the '*milieu interieur*' which is the condition of free and independent life," he wrote, and "all the vital mechanisms, however varied they may be, have only one object, that of preserving constant the conditions of life in the internal environment." "No more pregnant sentence," in the opinion of J. S. Haldane, "was ever framed by a physiologist."

Bernard, who emphasized especially the importance of freeing the organism from limitations set by the external world, listed water, oxygen, uniform temperature and nutrient supplies (including salts, fats and sugar) as the necessary constants. It is probable that we do not yet possess all the information needed for a complete list of the stabilizing factors, and a classification of those now known would be marred by cross-relations. We do know many of the factors. Sharp classification of them, however, is not essential to a discussion of their importance and the principles involved in their control. Obviously there are materials which must be provided as a source of the energy displayed in muscular movement, glandular secretion and in other activities, and also for growth and repair—such materials as glucose (grape sugar), protein (nitrogen-containing substances seen in meat, white of egg, etc.) and fat. There is also oxygen, there is water, and the inorganic salts, and finally there are "internal secretions," such as those coming from the thyroid and the pituitary glands, which have general and continuous effects. Furthermore, there are intimate environmental conditions which may profoundly affect cellular activity, such as the concentration of dissolved substances, the temperature, and the relative amounts of acid and alkali in the fluid matrix.

Each of the items mentioned above exists in a relatively uniform state in the internal environment of the living cells of higher organisms. There are oscillations, to be sure, but normally the oscillations are within narrow limits. If these limits are surpassed, very serious consequences may result, as we shall have many occasions to observe. Ordinarily the variations from the mean position do not reach the dangerous extremes which impair the functions of the cells or threaten the existence of the organism. Before those extremes are reached, agencies are automatically called into service which act to bring back towards the mean position the state which has been disturbed.

In later chapters we shall consider the ways in which these self-regulatory agencies operate to preserve constancy in the fluid matrix. Before coming to that phase of the exposition I propose to take up the action of certain agents which assure the primary, essential condition for cell life, that is, the maintenance and the effective use of the fluid matrix itself.

References

Bernard. Les Phénomènes de la Vie, Paris, 1878.
Haldane. Respiration, New Haven, 1922.

II

THE SAFE-GUARDING OF AN EFFECTIVE FLUID MATRIX

I

IN CONSIDERING the means employed to preserve the fluid matrix and to render it as useful as possible at critical times, we shall pay attention only to the blood, for the lymph may be regarded as secondary to that. In order that the blood shall continue to serve as a circulating medium, fulfilling the various functions of a common carrier of nutriment and waste, and assuring an optimum habitat for living elements, there must be provision for holding it back whenever there is danger of escape. The serious hemorrhage which occurs in "bleeders" (whose blood does not clot or clots too slowly) from so slight an operation as pulling a tooth, for example, is proof of that. I do not propose to describe the complicated changes which lead to the appearance of a clot, though the plug of blood jelly that forms over a cut and seals it is of fundamental value to the organism. Nor shall I emphasize the local contraction of injured blood vessels that lessens the possibilities of severe bleeding and makes easier the fixing of the seal. These well-known local conditions do not involve the more extensive physiological

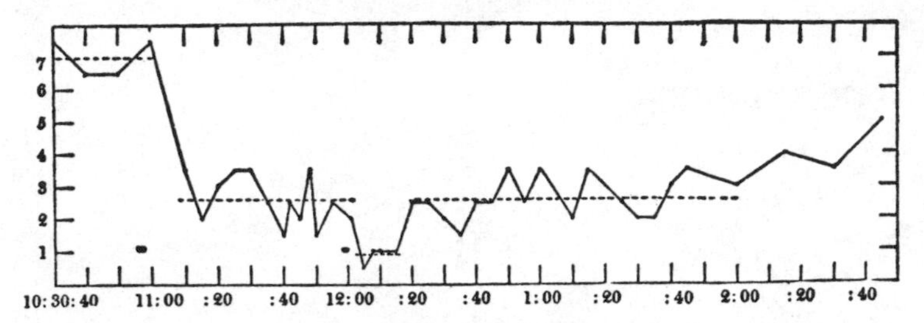

Fig. 2. Record showing shortening of coagulation time after a hemorrhage (13 per cent of the blood) at 10:59, and after a second hemorrhage (10 per cent of the blood) at 11:59. The dotted lines in this and the next figure indicate the averages for the time they cover.

reactions with which we are concerned. The aspect of the adjustment which takes place in hemorrhage, that is interesting in relation to homeostatic processes, is the increase in the speed of clotting as hemorrhage proceeds.

The phenomenon of accelerated clot formation as more blood is lost from the body has long been recognized. In the latter part of the eighteenth century Hewson noted that when an animal is bleeding to death, the latest blood which escapes clots more quickly than the earliest; and a century later Cohnheim reported that when an animal is killed by withdrawal of the blood in successive portions, the last portions sometimes coagulate almost instantaneously. In observations made by Gray and Lunt in the Harvard Physiological Laboratory this testimony was corroborated and amplified. As shown in figure 2, the coagulation time before the hemorrhage, in a typical experiment, as measured by an automatic device,[1] averaged about seven minutes. About 13 per cent of the estimated volume of the blood was

[1] The method is described in the volume "Bodily Changes in Pain, Hunger, Fear and Rage." 1929, p. 135.

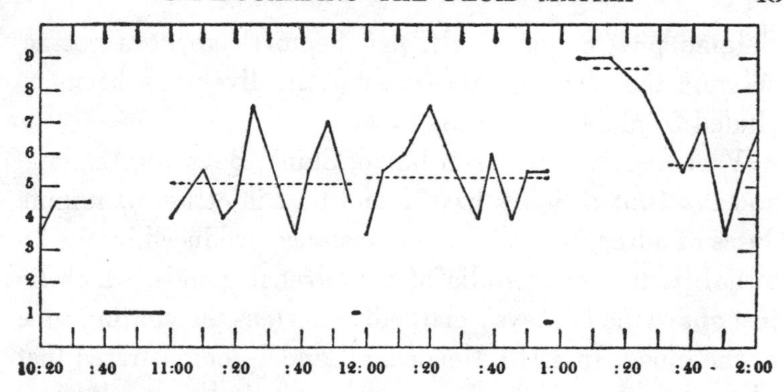

Fig. 3. Record showing absence of rapid clotting after hemorrhage, when the circulation is confined anterior to the diaphragm. From 10:40 to 10:58 the operation of tying the aorta and inferior cava above diaphragm was performed. At 11:58, 5 per cent of the blood was drawn, and at 12:58, 5 per cent again, each time with resulting respiratory distress.

then withdrawn (the animal, of course, was anesthetized). Note that the clotting time was reduced thereby to about two and a half minutes. Withdrawal of 10 per cent more reduced the time to about one minute. Thereafter, for a long period, about two and a half minutes were required, until there began a gradual restoration of the original rate.

Further observations showed that the more rapid coagulation attending hemorrhage does not occur if the circulation of the blood is confined to the front part of the animal. In figure 3 the long line represents the time which elapsed while the aorta and the inferior vena cava were being tied in the lower part of the chest. As revealed by the record, subsequent hemorrhages of 5 per cent of the estimated blood volume were not followed by faster clotting—indeed, the coagulation time was somewhat prolonged. This result was in harmony with the evidence brought forward by the

Belgian physiologist, Nolf, that hemorrhage does not accelerate the clotting process after the liver has been excluded from the circulation.

Various investigators in Japan, China, Belgium, England and the United States have found that injection of minute doses of adrenalin (the active substance produced by the internal portion or medulla of the adrenal glands, which lie just above the kidneys) markedly shortens the clotting time of the blood. In 1914 Mendenhall and I demonstrated that when the adrenal glands are made to discharge adrenin [1] into the blood stream, by stimulation of the splanchnic nerves of the sympathetic system (see fig. 35), the phenomenon of rapid clotting is induced. Figure 4 indicates that repeatedly, after such stimulation, the blood coagulates at a faster rate. It does not do so if the adrenal gland on the stimulated side has previously been removed. Furthermore, injected adrenalin has no accelerating effect on coagulation if the circulating blood does not enter the abdomen. It would seem, therefore, that the physiological method for hastening the clotting process requires the outpouring of adrenin into the blood stream and also the action of adrenin on some abdominal organ, probably the liver.

When hemorrhage occurs and arterial blood pressure falls, the sympathetic system is called into action. Tournade and Chabrol have proved that in these circumstances the adrenal medulla is stimulated to secrete adrenin. In other words, the conditions brought about by hemorrhage are such as to induce most effectively and in a wholly natural manner the more rapid coagulation which actually super-

[1] "Adrenin" is the natural secretion of the adrenal medulla; "adrenalin" is a commercial extract of the adrenal glands.

venes. From the experiments of Gray and Lunt, however, it appears that even after removal of the adrenal glands hemorrhage can still be attended by faster clotting. This result may be due to direct action on the liver cells of an

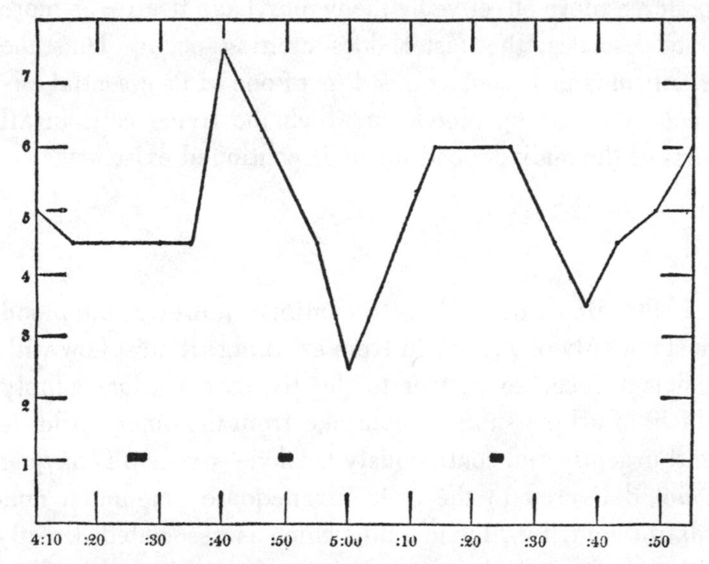

Fig. 4. Results of stimulating the left splanchnic nerves, 4.25–.28, after removal of the left adrenal gland, and of stimulating the right splanchnic nerves, 4.49–.51 and 5.23–.25, with right adrenal gland present.

inadequate supply of blood, for there is evidence that these cells are peculiarly sensitive to oxygen want. The fact that hemorrhage *can* cause acceleration of the clotting process, although the adrenal glands are absent, does not, however, minimize the importance of the rôle of the sympathico-adrenal system in that process under normal conditions; we shall come upon numerous instances of compensatory arrangements in the organization of bodily processes. In the usual course of events the stimulation of the sympathico-adrenal system by hemorrhage would automatically result in prompt clotting of the escaping blood. That is, the natural conservative agency, coagulation, would automatically be rendered more effective by the hemorrhage itself—as more blood escapes, the faster does clotting occur. Thus the organism is protected against loss of one of its essential elements, the flowing blood, on which the living cells in all parts of the body depend for their continued existence.

II

If the life of the cells is to continue, however, the blood must not only be prevented from escaping; it must flow with sufficient speed to deliver to the living parts that supply which, of all the supplies obtained from the outer world, is most urgently and continuously needed—oxygen. If oxygen is not delivered to the cells in adequate amount, a non-volatile acid, i.e., lactic acid, which is associated directly with cellular activity (e.g., contraction of muscle cells), is not burned to the volatile carbonic acid which is readily carried away by the blood to the lungs and there breathed out. If lactic acid accumulates in the cells it soon diffuses out into the surrounding fluids. In the blood it unites with the alkali, sodium bicarbonate, and changes it to sodium lactate and carbonic acid, which is discharged in the usual way. The details of this change we shall have to examine later (see Chapter XI). At present I wish merely to mention that by the change from sodium bicarbonate (the "alkali reserve" of the blood) to sodium lactate, the alkali reserve is reduced, and that the reduction can be used to indicate that the supply of oxygen to the tissues is not sufficient to burn the non-volatile acid which is constantly being produced.

Experiments which McKeen Cattell and I performed at Dijon during the World War brought out evidence that

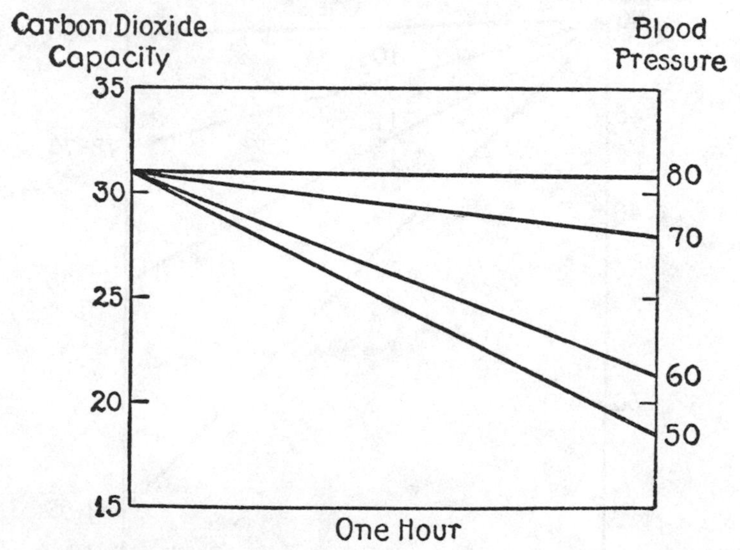

Fig. 5. Progressively greater reduction of the alkali reserve (as measured by the capacity of the blood plasma for carbon dioxide) as the blood pressure is reduced below 80 millimeters of mercury; e. g., reduction to 60 millimeters for one hour lowered the carbon dioxide capacity from 31 to 18.5 volumes per cent.

there is a critical level in a falling arterial blood pressure, at which signs appear that the volume of blood that is flowing in a given time is becoming inadequate. The alkali reserve can be measured by the carbon dioxide content of the blood plasma which has been exposed to a standard concentration of that gas. As shown in figure 5, reduction of the arterial pressure from about 120 millimeters of mercury to 80, for an hour, had no effect on the alkali reserve; reduction to 70, 60 or 50 millimeters for the same

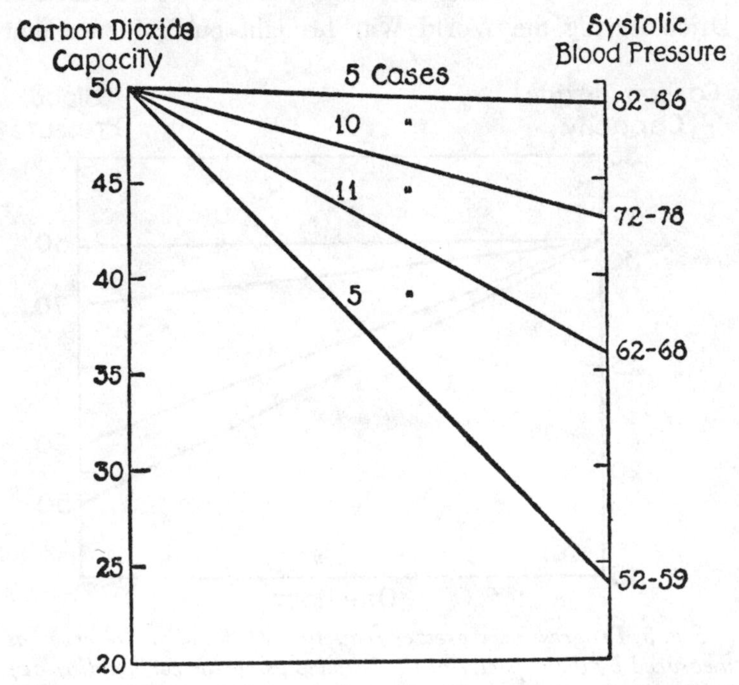

Fig. 6. Relation of the carbon dioxide capacity of the blood plasma to systolic blood pressure in 31 cases of shock and hemorrhage in wounded men.

time had progressively greater effects in lessening the sodium bicarbonate of the plasma. According to the interpretation outlined above the critical level in a falling arterial blood pressure is approximately 80 millimeters of mercury. Below that level the pressure is not capable of maintaining a volume-flow, to organs which must continue active, that is sufficient to serve their normal oxidative processes.

Records of wounded men tell the same story as the experiments just described. In figure 6 are presented the results of observations on 43 such cases which Fraser, Hooper and I studied at Béthune in 1917. They are reported on the basis of the systolic blood pressure. As the figure shows, the alkali reserve rarely falls below the normal for man (which is represented by about 50 per cent of carbon dioxide by volume) until the systolic pressure is less than 80 millimeters of mercury. Furthermore, just as in the experiments performed at Dijon, the greater the reduction of the blood pressure below that critical level, the greater is the fall in the alkali reserve.

The view that a blood pressure below a critical level is actually unable to furnish enough blood to meet the requirements of active organs is confirmed by various observations. Markwalder and Starling found that the contractions of the isolated heart soon begin to weaken if the blood pressure is held for some time below 80 or 90 millimeters of mercury. And a number of investigators in England and the United States have shown that prolonged low pressure results in such injury to that part of the sympathetic nervous system that controls the muscles in the arterioles (the vasomotor nerves) that all possibility of obtaining reflex responses is lost. Figure 7 presents a copy of an original record in an experiment performed at Dijon in 1918. By regulated pressure around the heart (in an anesthetized animal) the arterial pressure could be set at any level and kept there. It was held at 60 millimeters of mercury for an hour. When release came the arterial pressure promptly rose to nearly its original height. But as the pressure was kept low, hour after hour, the vasomotor system lost its resiliency, and at the end of three hours it was incapable of any restorative action. When we consider

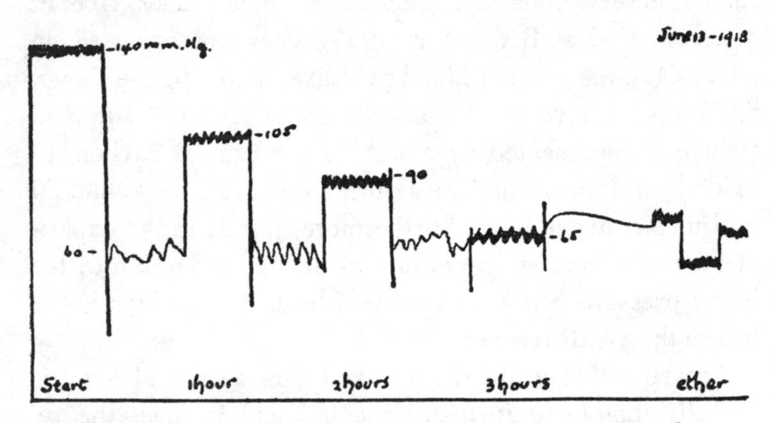

Fig. 7. Gradual failure of the blood pressure to rise as the pressure is held at 60 millimeters of mercury for successive periods of one hour and then released for five-minute intervals.

the special sensitiveness of nerve cells to even partial anemia, we can understand how readily a diminished volume-flow of blood might have injurious effects.

Confirmatory of the view that active organs suffer from oxygen-want, if the blood pressure is not sustained above a critical level, are the observations of one of my associates, Joseph Aub. He studied the basal metabolism (i.e., the total oxidative processes of the organism under standard conditions) in cases of experimental shock with low blood pressure and found an average reduction of 18.5 per cent in eight cases of mild shock (e.g., with a systolic blood pressure of about 70 millimeters), and an average reduction of 33 per cent in eight cases of severe shock (with a blood pressure near 60 millimeters). By holding the level of blood pressure artificially at about 60 millimeters Aub demonstrated that the metabolism was reduced approximately 30 per cent; in other words, the low metabolism was due to the lessened blood flow and not to the procedures which were used to induce shock.

III

In the foregoing paragraphs I have described in some detail examples of the failure of the circulating blood to perform its most urgently important function as a common carrier—the delivery of sufficient oxygen to supply the minimal needs of the body—because I wished to lay emphasis on the devices which are employed in the organism to avoid such conditions and their bad consequences.

First of all, when hemorrhage occurs, the sympathetic system is brought into action. That was demonstrated by Pilcher and Sollman in 1914, when they observed that the blood vessels of an organ which was arranged for special examination responded to hemorrhage by contracting—an effect produced through vasomotor nerves. This evidence was later confirmed by Bayliss and Bainbridge. The same phenomenon occurs when the blood pressure falls during the development of shock. In figure 8 are shown graphs plotted from the averages of six experiments performed by one of my colleagues, McKeen Cattell. To test the contraction of the blood vessels he recorded the perfusion time, i.e., the time required for a standard amount of physiological salt solution (0.9 per cent sodium chloride), under

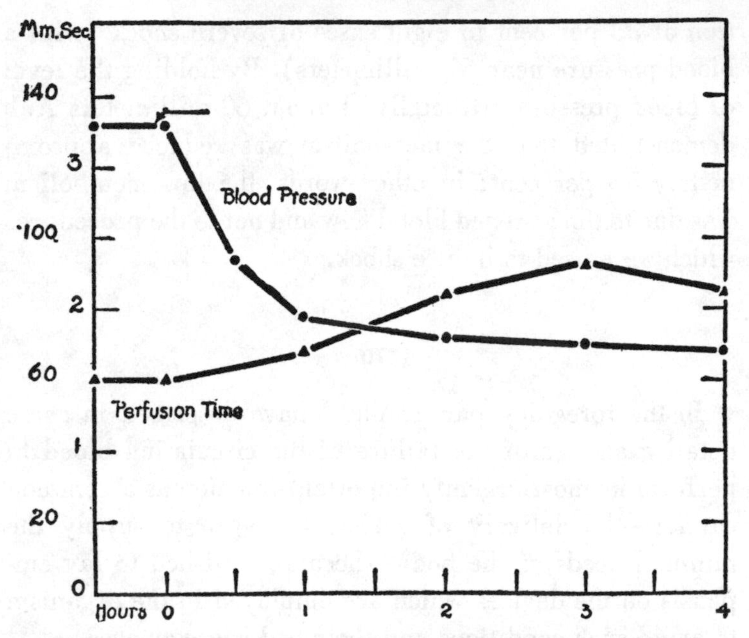

Fig. 8. Graph plotted from the averages of six experiments showing the relation between the perfusion rate and the blood pressure in shock caused by muscle injury at the time indicated by the arrow. As the blood pressure falls, there is a slowing of the perfusion rate, indicating an increased tone of the arterioles.

standard pressure, to pass through the arterioles of the leg. If the vessels should contract they would present greater resistance to the passage of the solution and the time would lengthen. Note (in fig. 8) that as the blood pressure fell the perfusion time did in fact gradually lengthen. At the end of three hours it had increased more than 60 per cent. Up to that stage the reduction of the blood pressure had been met by action of the vasomotor center in the brain, action which caused constriction of the peripheral arterioles.

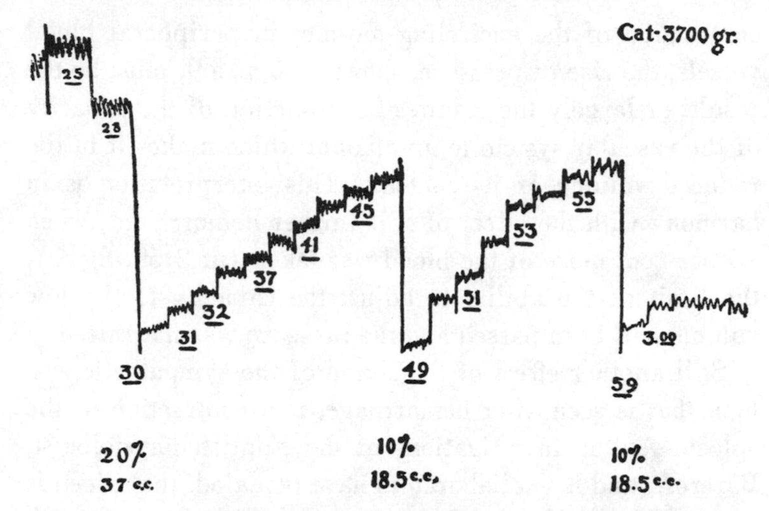

Fig. 9. A record showing the prompt recovery of blood pressure after hemorrhage. The recovery is due to a constriction of the blood vessels that adjusts their capacity to their contents. The limit of the adjustment was reached when, after 30 per cent of the blood had been removed, 10 per cent more was withdrawn. (*From Cannon: "Traumatic Shock." Courtesy of D. Appleton and Co.*)

A graphic picture of the effects of peripheral vasoconstriction is shown in figure 9, the record of an experiment performed at Dijon during the War. At 2:30 o'clock 20 per cent of the estimated blood volume of an animal was withdrawn. The arterial blood pressure, as shown by the record, fell precipitately. In fifteen minutes, however, it had been restored nearly to the former level. Then 10 per cent more of the blood was removed. In about six minutes the pressure was again restored. Examination of the blood proved that the restoration was not due to increase of blood volume by inflow of lymph from the tissue spaces, for the coloring matter of the blood (hemoglobin) was not diluted. Since bleeding, and consequent fall of blood pressure, is followed by contraction of the encircling muscles in peripheral blood vessels, the rise of pressure, shown in figure 9, must be the result, or largely the result, of a reduction of the capacity of the vascular system to an amount which makes it fit the reduced volume of its contents. This interpretation is in harmony with the effect of still further hemorrhage; when 10 per cent more of the blood was taken (at 2:59, fig. 9), the limit of the ability to adjust the capacity to the low volume had been passed and the pressure was not raised.

Still another effect of the action of the sympathetic system, that is seen after hemorrhage, is a contraction of the spleen. As the investigations of the English physiologist, Barcroft, and his collaborators have revealed, the spleen is a reservoir of blood, where the red corpuscles are much concentrated. When severe bleeding occurs the lost blood is at first progressively compensated for by contraction of the spleen. The surface of that organ, before and after hemorrhage (see fig. 10, from an article by Barcroft), indicates how greatly the organ can contract, but the reader should interpret the outlines with remembrance that the spleen has *thickness* as well as length and breadth! The concentrated blood, thus added to the circulation, protects the organism against the possibility of disturbance by losses in the early stage of hemorrhage, and if the hemorrhage is checked, may satisfactorily compensate for the losses.

Now, what is the significance of all these adaptations and adjustments when the common carrier and rapidly mobile agent in the fluid matrix, the blood, escapes from its courses, so that the effective head of pressure, which keeps it moving at its proper pace, is lost? In order to understand the meaning of these changes we must realize that certain

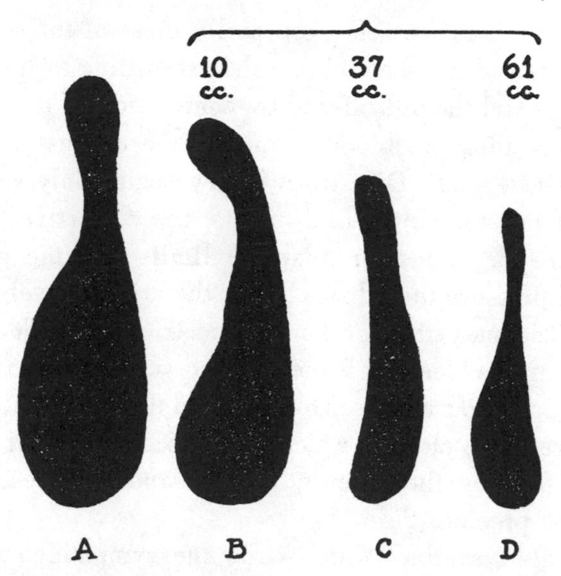

Fig. 10. Surface of the spleen of a cat. A with the animal under urethane anesthesia, B after loss of 10 cc. of blood, C after loss of 47 cc., and D after loss of 108 cc., when the animal died. (Records taken by Barcroft.)

structures in the body are essential to continued existence—the brain with its control of respiration and swallowing, and the heart and the diaphragm. There is evidence that the extraordinarily sensitive brain and the continually active cardiac muscle have a volume-flow of blood through their vessels that is dependent directly on the general arterial pressure. If the pressure falls below the critical level, as we have seen, these organs may suffer irreparable damage. The constriction of the blood vessels that shuts down on the volume-flow elsewhere does not operate in these organs if the general arterial pressure is kept high. Even though there is a large hemorrhage, therefore, the vasoconstriction of the peripheral vessels—especially those of the skin, the fat depots and the skeletal muscles, according to Rous and Gilding—and the aid offered by contraction of the spleen, assure an adequate blood supply to the organs on which life itself depends. Disastrous injury begins only when, in spite of the automatic adjustments, the corrective devices are taxed beyond their adaptive limits and the general arterial pressure then drops below the critical level.

An incidental effect of the vasoconstriction which characterizes serious bleeding is the reduced volume-flow of blood in the peripheral vessels. This lessened flow coöperates with the more rapid clotting which is induced by loss of blood, so that together they prevent a large and rapid escape of this most precious fluid.

The alert sentinel which calls the sympathico-adrenal system into service when dangerous blood-loss is threatened appears to be some sensitive nerve endings in the blood vessels high in the neck, near the brain. The two large arteries on either side of the windpipe (the carotid arteries) branch in that region, the inner branch leading to the cerebral vessels. Where the branching occurs there is a bulbous enlargement, called the "carotid sinus," which is especially supplied with nerves. In 1910, the French physiologist, Hédon, demonstrated that when the blood pressure falls in the arteries of the head, there is general constriction of the blood vessels of the rest of the body. These observations were concordant with a previous observation by Porter and Pratt that the pressure in the blood vessels of the foot is reciprocally related to the pressure in the carotid artery. Tournade, Chabrol and Marchand, and also Anrep and Starling, have confirmed this testimony. It was the

Belgian physiologist, C. Heymans, who refined the evidence and proved that the low blood pressure in the head region acts as a stimulus to the nerve endings in the carotid sinus. The reflexes thus aroused are displayed in the action of the sympathetic nervous system which causes widespread constriction of blood vessels, contraction of the spleen, and discharge of adrenin from the adrenal medulla. Thus all of the phenomena which we have been considering as adaptive and corrective, and tending to preserve the welfare of the organism by maintaining an effective use of its fluid matrix, can be explained as automatic consequences of the lowered arterial pressure due to blood-loss.

IV

Although there is compensation for hemorrhage by general vasoconstriction we must remember that it is at best a temporary makeshift. The brain and heart, to be sure, are supplied with blood, and that is of primary importance, but peripheral organs, resident in regions where vessels are constricted, are not receiving their normal supply. The only efficient way to meet that situation is to increase the volume of blood until it properly fills the usual capacity of the vascular system. Within limits an increase in the volume of the circulating blood is brought about by the passage of water (and salts) from the lymph in the tissue spaces through the walls of the capillaries into the blood stream. This process is explained as due to a lessening of the filtration pressure (which presses water through the capillary walls into the lymph) when loss of blood reduces the pressure in the capillaries, and also to the tendency of the water of the lymph to diffuse back into the blood because the concentration of water in blood is less than in lymph. (Or, to put the situation in other words, there is more colloid material in blood plasma than in lymph; the osmotic pressure of blood is therefore higher than that of lymph; and consequently water and salts pass from the lymph to the plasma.) As just indicated, this process becomes effective as the capillary pressure falls. Thus the rapidly circulating portion of the fluid matrix, the blood, receives support from the more slowly moving portion, the lymph, and gradually, with the taking of more fluid into the body, the volume of the blood is restored. The restoration of the normal number of red corpuscles is a considerably slower process.

The removal of water and salts from the peripheral lymph, or the failure to supply these regions with the water they need, results in a number of interesting effects, one of which is the phenomenon of thirst. After a battle the universal cry of men who have been badly wounded and who are suffering from hemorrhage or shock, is the cry for water. Often they are, unfortunately, unable to retain what they drink. When water is taken and absorbed, however, it may be called for in astonishingly large amounts. Water which enters the body by way of the digestive canal, as shown by Robertson and Bock, is much more effective in restoring a reduced blood volume than is physiological salt solution injected into a vein or under the skin. Thirst, therefore, is normally an accurate and sensitive indicator not only of the body's need for water, but also of the best mode of supplying water to the organism.

V

We have reviewed the evidence that the freedom and independence of our existence, in the presence of profoundly disturbing conditions either in the outer world or in our own organization, are dependent on the existence and constancy of a fluid matrix in which our living parts reside. We have seen that in our bodies there are devices which are ready to operate so that, whenever danger of loss of the fluid matrix appears, agencies are quickly brought into action which lessen the danger. The blood is made to clot faster and its volume is reduced by continued escape from the body. The peripheral blood vessels are constricted; thereby not only is the flow reduced in regions where loss of blood is most likely to occur, but also a persistent supply is assured to essential and sensitive organs, the brain and the heart. All these adjustments are managed automatically by the sympathetic system, roused to action by the lowered blood pressure. This emergency adjustment in the organism is followed by restoration of blood volume from supplies of water and salts in the tissue spaces, and from the functional service of thirst. In the next chapter we shall see how thirst itself, as well as hunger, may be explained as means of providing the supplies needed for homeostasis in the internal environment.

References

Anrep and Starling. Proc. Roy. Soc., 1925, xcvii, 436.
Aub. Am. Journ. Physiol., 1920, liv, 388.
Barcroft. The Lancet, 1925, i, 319.

Bayliss. The Vasomotor System. London, 1923.
Cannon and Cattell. Arch. of Surgery, 1922, iv, 300.
Cannon, Fraser and Hooper. Journ. Am. Med. Ass'n., 1918, lxx, 607.
Cannon and Mendenhall. Am. Journ. Physiol., 1914, xxxiv, 243.
Cohnheim. Allgemeine Pathologie, 1877; or Cohnheim's "General Pathology," London, 1889.
Gray and Lunt. Am. Journ. Physiol., 1914, xxxiv, 332.
Hédon. Arch. internat. de Physiol., 1910, x, 122.
Hewson. An Experimental Inquiry into the Properties of the Blood, London, 1772.
Heymans. Le Sinus Carotidien. Louvain and Paris, 1929.
Markwalder and Starling. Journ. Physiol., 1913, xlvii, 279.
Nolf. Arch. internat. de Physiol., 1905–6, iii, 1.
Pilcher and Sollman. Am. Journ. Physiol., 1914, xxxv, 59.
Porter and Pratt. Am. Journ. Physiol., 1908, xxi, p. xvi.
Robertson and Bock. Rep. of Shock Com., British Med. Research Com., 1918, No. 25, 135.
Rous and Gilding. Proc. Soc. Exper. Biol. and Med., 1929, xxvi, 497.
Tournade and Chabrol. C. r. Soc. de Biol., 1925, xciii, 934.
Tournade, Chabrol and Marchand. C. r. Soc. de Biol., 1921, lxxxiv, 610.

III

THIRST AND HUNGER AS MEANS OF ASSURING SUPPLIES

I

In the volume, "Bodily Changes in Pain, Hunger, Fear and Rage," I have discussed in detail the nature of thirst and hunger. The emphasis there, however, was on these phenomena as motives for action, as "drives." They are of primary value, of course, in making certain that the organism has the essential materials, water and food, for carrying on its functions, and in that relation we shall consider them here.

We all know that water and food are among the fundamental requirements of the organism and that at all times both water and food are being spent. Non-volatile waste is constantly passing out through the kidneys. And volatile waste from the burning of food material in the tissues is carried away by every breath. Acting as a vehicle for the discharge of this waste is water. Also water is continuously being lost by evaporation from respiratory and cutaneous surfaces.

Since water and food are being steadily lost from the body, the only way in which a constant supply can be maintained is by means of storage and gradual release. Water is stored in tissue spaces and in tissue cells. Food is stored in the well-known form of fat, and as animal starch or glycogen and probably also as protein in small masses in the cells of the liver. When need arises these stored reserves are set free for use. The reserves themselves, however, must be replenished. It is the function of thirst and hunger as automatic stimuli to make certain that the reserves of water and food are maintained.

II

First, let us consider thirst as a means of assuring a proper water supply. Thirst is a sensation referred to the inner surface of the mouth and throat, especially to the root of the tongue and to the back part of the palate—a highly unpleasant sensation of dryness or stickiness. It is commonly associated with experiences which result in a rapid evaporation of the moisture of the mouth or a diminished production of fluid there. For instance, the breathing of hot, dry air, as well as prolonged speaking or singing, or the chewing of desiccated food, will lead to the sensation of thirst and the desire for something to drink. Fear and anxiety likewise are associated with dryness of the buccal mucous membrane and may cause distressing thirst.

Besides these local conditions, however, there are certain general bodily states which may produce the sensation. Profuse sweating, for example, or the excessive loss of fluid from the body in the course of disease, as in the diarrhea of cholera or in the abundant loss of water through the kidneys in diabetes, will provoke thirst to an intense degree. And, as we noted in the last chapter, after severe hemorrhage thirst is tormenting.

In accordance with the evidence that thirst is due to local dryness of the mouth and also that it accompanies states of water need in the body as a whole, there have developed two sets of theories concerning the nature of the sensation; some experimenters have advocated the view that thirst is of local, peripheral origin, and others have argued that it is a general sensation.

The evidence supporting the idea that thirst has a general or diffuse origin is in the main derived from certain general modes of treatment of thirst which affect the body as a whole and which abolish the sensation. For example, thirst may be promptly banished by the injection of water under the skin or into the intestines. The introduction of water by these routes does not moisten the pharynx and yet the desire for water disappears.

The main fact which must be kept in mind, however, is that the thirsty man does not complain of a vague general condition; he complains of a parched and burning throat. And there are other facts which point to a local origin. Persons suffering from severe thirst, because of great losses of water through the kidneys, are freed from their distress when the sensitiveness of the nerve endings at the back of the mouth is destroyed by cocaine. Moreover, sipping a small amount of water and moving it about in the mouth will stop the sensation. Also, holding on the tongue a substance which causes secretion of saliva—a bit of lemon, for instance—will lessen thirst. None of these procedures supplies water to the body, and yet the distress is mitigated. Their efficacy in bringing relief does not suggest, however, any rational account of the relation between the local dryness and the water lack in the organism as a whole. We should understand how the dry mouth as a local condition may be a means of automatically indicating bodily need and of automatically leading to satisfaction. Clearly, an arrangement must be sought which, when the body requires water, would cause the mouth to become dry. This arrangement might peculiarly be expected to be developed in animals which are continuously and rapidly losing water and which therefore must have repeated renewal of the water supply in order to maintain a normal condition. Let us follow these clues and see where they lead us.

Water-inhabiting animals, e.g., fish, with a wet integument and with water passing through the mouth and out through the gills, perhaps do not experience thirst. Animals surrounded by air, on the other hand, have a dry integument in contact with the air; and, as shown in figure 11, instead of a water current which passes through the mouth, there is an air current which passes through the nose and across the ancient water course. The nose and trachea are well provided with moistening glands, but at the crossing there are few such glands. Therefore prolonged speaking, singing or smoking, in which air is drawn through the mouth and not through the nose, tends to dry this region, and the sensation of dryness and stickiness there, as we have noted, is commonly recognized as thirst.

But why does not the region always feel dry and sticky? And why does it feel so when the body needs water? Again, if we compare the water animals with the land animals, we find that only the land animals have special buccal glands—in higher forms the salivary glands.

The theory which these facts suggest is that when water

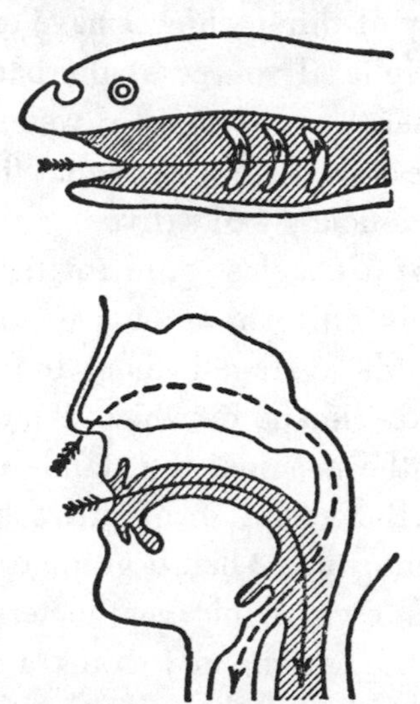

Fig. 11. Midsection of the head of a fish and the head of a man to show the new relations of the water course from the mouth to the gill region in air-inhabiting animals. Note that in the pharynx the air current passes to and fro across the ancient water course (which is cross-hatched). (From Cannon: Article "Hunger and Thirst," "Foundations of Experimental Psychology." Courtesy of the Clark University Press.)

is lacking in the body the salivary glands are unfavorably affected, along with other structures, by the deficient water supply; that they differ from other structures, such as muscle, in that they require a large amount of water for the performance of their function, which is that of pouring out a secretion consisting almost wholly of water: and furthermore, that they occupy a peculiarly strategic position, for if they do not have water to apply in their secretion and are therefore unable to secrete, the mouth and pharynx become dry and thus the sensation of thirst arises. Such is the theory of thirst which I have offered to account for its local, peripheral source at the back of the mouth, when there is general bodily need of water.

Let us examine now the evidence that thirst results from a dryness due to deficiency of saliva.

The chewing of a tasteless gum for five minutes evokes repeatedly a fairly uniform amount of saliva. In observations on myself this averaged about fourteen cubic centimeters. In an experiment in which I took no fluid after seven o'clock in the evening, the standard mastication resulted in no smaller secretion of saliva until after eleven o'clock the next morning. Then it gradually fell from fourteen to less than eight cubic centimeters. Thereupon, at three o'clock in the afternoon I drank a liter of water. A collection of saliva during the following four hours, by the standard method, proved that the discharge from the salivary glands had been promptly restored to approximately the previous amount and was there maintained. One of my colleagues, Magnus Gregersen, has discovered that when a dog is placed in a warm room of standard temperature, the panting is accompanied by a remarkably uniform outflow of saliva from one of the salivary glands at the base of the tongue (the submaxillary gland), the product of which, by appropriate means, can be collected in a measuring tube. As shown in figure 12, deprivation of water greatly reduces the dog's salivary secretion; and drinking quickly restores the secretion to its usual amount. This observation confirms the results which I had noted and proves that they were not vitiated by disturbing subjective states. In the experiments on myself the sensation of thirst became prominent during the period when the saliva flow was diminishing. After the water was drunk and the saliva was again

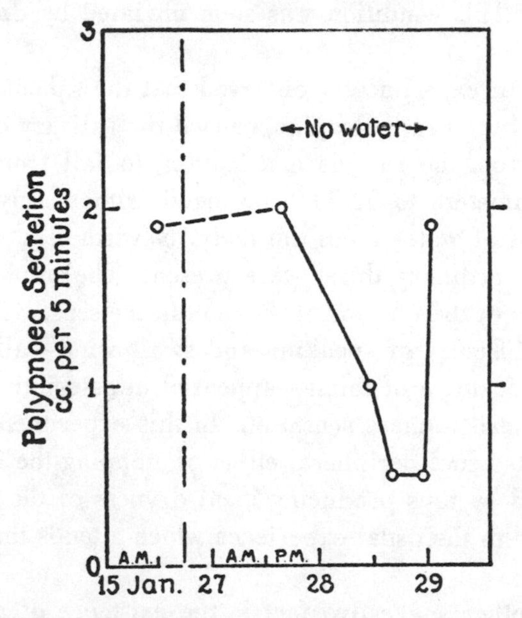

Fig. 12. The effect of water deprivation on salivary secretion due to panting. The secretion was promptly restored to its normal rate when the dog was given water.

secreted the sensation disappeared. The coincident association of bodily need for water, diminished flow of saliva, and the sensation of thirst, strongly suggests that the deficient functioning of the salivary glands signals the bodily need by causing the unpleasant sensation.

Again, by wrapping the body in very warm blankets and applying hot water bottles it is possible to cause abundant sweating. As a consequence of this large loss of water from the body, the output of saliva, due to chewing for a standard period, was diminished to about one half what it was before the water loss. Associated with the lessened salivary flow was a noteworthy dryness of the mouth and an unpleasant thirstiness. This condition was soon obviated by drinking water.

In another experiment I observed that the subcutaneous injection of the drug, atropine, caused the salivary output, resulting from the routine mastication, to fall from 13.5 cubic centimeters to 1. This occurred without any noteworthy loss of water from the body. Nevertheless, all the feelings of ordinary thirst were present. The unpleasant, dry surface of the interior of the mouth, the sense of stickiness, the difficulty of speaking and swallowing—all these particular features of thirst—appeared in association with the characteristic mass sensation. In this experiment atropine had its usual peripheral effect of stopping the flow of saliva; and by thus producing local dryness of the mouth it gave rise to the usual experience which attends that condition.

Still another suggestive fact is the existence of a well-established reflex secretion of saliva after the mouth becomes slightly dry. A very simple experiment is that of chewing an indifferent substance for five minutes and comparing the amount of saliva thus obtained with that which results from breathing only through the mouth for five minutes. At first the passage of air to and fro through the mouth gradually dries the surface. As the surface becomes more and more dry, however, saliva is poured out, and the amount collected as it flows forth may be, in my experience, considerably greater than that obtained by the chewing. The presence of this reflex indicates that salivary glands have as one of their special functions the moistening of the mouth.

Thirst as a consequence of fright, and the attendant checking of salivary secretion, are well-known phenomena. Dr. H. J. Howard has vividly reported his experiences when he thought he was about to be shot by Chinese bandits. "So I was going to be shot like a dog!" he wrote. "My tongue began to swell, and my mouth to get dry. This thirst rapidly became worse until my tongue clove to the roof of my mouth, and I could scarcely get my breath. The thirst was choking me. . . . I was in a terrible state of fear." He prayed for strength to meet his approaching doom and soon fear left him as he determined to die like a man. "Instantly my thirst began to disappear" he testified. "In less than a minute it was entirely gone and by the time we had reached the gate I was perfectly calm and unafraid." Note that the intense and distressing sensation of thirst was not associated with a real lack of fluid in the body but resulted from a local condition in the mouth.

The foregoing observations taken together support the conclusion that thirst is normally the consequence of a drying of the mucous membrane of the mouth and pharynx when the salivary glands fail to keep this region moist. The continuous loss of water from the body through the kidneys, through the respiratory passages and the skin, does not for a long period cause any appreciable change in the water content of the *blood*. Observations made by the French physiologist, André Mayer, have shown that there may be no demonstrable change in the blood of a dog after three days of total deprivation of water. The blood as the active part of the fluid matrix is maintained in a constant state at the expense of the water reserves in the tissues and in the cells of the body structures. Among the structures which are called upon are the salivary glands. As pointed out above, however, they require water for their proper service to the organism. Not having water available they cannot perform that service and consequently the mouth becomes disagreeably dry. When water is drunk it is at once made available for the salivary glands, as well as other organs, and they can again perform their special function of keeping the mouth parts moist and lubricated. These glands serve, therefore, as indicators of bodily need for water.

III

Now let us turn our attention to hunger as a means of assuring food supplies. Hunger has been described as a very disagreeable ache or pang or sense of gnawing or pressure which is referred to the epigastrium. The older theories of hunger assumed that it was a "general sensation" which was based on a widespread need for food in the organism. The lack of food in the circulating blood was supposed to stimulate the cells of the brain, among others, and thus the sensation was explained. According to this view, the common reference of the sensation to the vicinity of the stomach is due to an association developed by the disappearance of the hunger pang when food is taken into the stomach. There are, however, serious objections to that view, which I have considered fully elsewhere.

The observation of fairly frequent and rhythmic recurrence of hunger pangs, together with associated sounds of moving air in the stomach region, led me, in 1911, to express the view that the pangs might be due to strong periodic contractions of the muscle of the gastric wall. Shortly thereafter opportunity to test the correctness of this idea was offered by one of my students, Washburn, who accustomed himself to the presence of a rubber balloon in his stomach and a small tube in his esophagus. The tube led from the gastric balloon to a recording apparatus, as shown in figure 13. A hollow elastic cylinder (a pneumograph) around the abdomen provided a means by which the move-

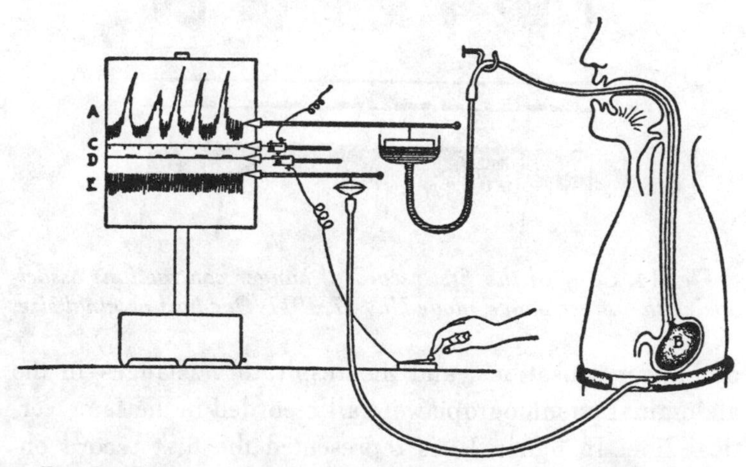

Fig. 13. Diagram showing the method used to record the gastric hunger contractions. A, Kymograph record of the increase and decrease of volume of the gastric balloon, B. C, time record in minutes. D, record of the subjective experience of hunger pangs. E, record of the pneumograph placed about the waist; this record proves that the hunger contractions do not result from action of the muscles of the abdominal wall. (*From Cannon, Article "Hunger and Thirst," "Foundations of Experimental Psychology." Courtesy of the Clark University Press.*)

ments of respiration were registered and allowed us to conclude that the pressure changes recorded from the stomach were not due to contractions of the abdominal muscles. A key in the subject's right hand was pressed when he felt the sensation of hunger. The volume changes in the gastric balloon, the time marked in minutes, the subjective testimony

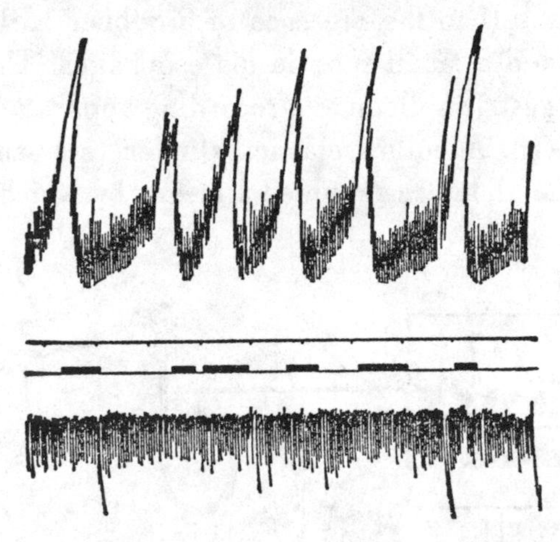

Fig. 14. Copy of the first record of hunger contractions associated with hunger pangs, made May 17, 1911. One-half original size.

of hunger sensations, and the respiratory changes in the abdominal pneumograph were all recorded in the same vertical line. In figure 14 is represented the first record obtained by use of this method.

As shown in figure 14 there are powerful periodic contractions of the empty stomach, lasting approximately 30 seconds and recurring at intervals which vary from 30 to 90 seconds, with an average interval of about 60 seconds. As you will observe, the testimony of the subject that he experienced a hunger pang was not recorded until the contraction had nearly reached its peak. The sensation therefore was not the cause of the contraction—the contraction was the cause of the sensation.

The observations made by us were soon confirmed by Carlson, who studied the phenomena of hunger in a man with a gastric fistula, and also in himself by becoming accustomed to a balloon in his stomach as Washburn had done. In a series of interesting researches on human beings and on various kinds of lower animals, Carlson and his students have brought out many new aspects of the hunger pangs in relation to gastric contractions. They have shown that hunger usually begins with occasional weak contractions of the empty stomach, and that these contractions become gradually more vigorous and appear at shorter and shorter intervals until an acme of activity is reached which may end in a true spasm of the gastric muscle. Either the single contractions or the spasm may be associated with the typical ache or pang or gnawing sensation which has long been recognized as the experience of hunger. After the acme of activity has been reached the stomach usually relaxes and remains inactive for a period, whereupon it starts again with the occasional, weak contractions, and the cycle which has just been described is repeated.

The puzzling problem arises as to what induces the stomach, while empty, to contract with a vigor which is greater than that observed in the ordinary traveling contractions (peristalsis) which recur regularly during the digestion of a meal. It is known that the elective source of energy for muscular contraction is carbohydrate food—glycogen or sugar. It seemed possible that a deficiency of this energy-yielding material might be signalized by excessive contractions of the smooth muscle of the stomach. Bulatao and Carlson found that if the sugar concentration of the blood was reduced about twenty-five per cent by use of insulin, the hunger contractions became more intense—an observation which has been noted subjectively in human beings who have received an overdose of insulin. On the injection of sugar into the blood stream hunger contractions were abolished. Quigley and Carlson have confirmed the observation that insulin induces increased gastric (and duodenal) motility and corresponding hunger sensations and have found that the phenomena are promptly inhibited by introducing glucose into the duodenum where it can be absorbed. Since atropine, given subcutaneously, likewise abolishes the phenomena, it appears that the hunger contractions are caused by hypoglycemia acting through vagal influences. On gradually lowering blood sugar by removal of the liver La Barre and Destrée observed that gastric contractions began to appear when the glycemic percentage reached about 75 mg. per cent (see p. 98), and that their intensity and frequency increased as the percentage dropped to lower levels. This influence, however, was limited. At 45 mg. per cent, when convulsions occurred, the stomach was relaxed between the convulsive seizures. Likewise Quigley and Halloran noted that during the secondary drop of blood sugar below the normal level, after an intravenous injection of glucose, hunger contractions appeared, and that when the glycemic percentage rose again, they stopped. These observations seem to indicate that hypoglycemia alone, exclusive of the action of insulin, is associated with the occurrence of vigorous activity of the gastric muscles. When, however, Scott, Scott and Luckhardt examined in man the glycemic percentage in relation to the incidence of the natural hunger contractions, they found that it continued practically unchanged for long periods during which the typical contractions came and went. The explanation, therefore, of this peculiar behavior of the stomach, which originates the pangs of hunger, remains to be discovered.

IV

The ways in which appetites for food and drink and sensations of hunger and thirst act to maintain the bodily supplies of nutriment and water may be regarded as typical of other arrangements in the organism which operate for the welfare of the individual or the race. Behavior may be directed either by movements to get rid of disturbing, annoying stimulation, or by movements to prolong or renew agreeable stimulation. Hunger and thirst belong to the first category. Each of these states is associated with a natural impulse; each one more or less vigorously spurs or drives to action; each may be so disturbing as to force the person who is afflicted to seek relief from intolerable annoyance or distress. On the other hand, experience may condition behavior by revealing that a certain food or drink is the cause of unanticipated delight. An appetite for the repetition of this experience is thus established; the person beset by an appetite is tempted, not driven, to action—he seeks satisfaction, not relief. It is not to be supposed that the two motivating agencies—the pang and the pleasure—are as separate as we have been regarding them for purposes of analysis in the present discussion. They may be closely mingled; when relief from hunger or thirst is found, the appetite may simultaneously be satiated. Insofar as an assurance of supplies of food and water is concerned, appetite, or the habitual taking of these substances, is the prime effective agency. If the requirements of the body are not met, however, in this mild and incidental manner, hunger pangs and thirst arise as powerful, persistent and tormenting stimuli which imperiously demand the ingestion of food and water before they will cease their goading. By these automatic mechanisms the necessary supplies for storage of food and water are made certain.

Coöperating with hunger and thirst in a way not yet clearly defined is the sensation of having had enough. Protection of the organism against being overstocked with food and water is thus obtained. The feeling of satiation is little understood, but it is important and deserves further attention.

References

Bulatao and Carlson. Am. Journ. Physiol., 1924, lxix, 107.
Cannon. Thirst. Proc. Roy. Soc., 1918, xc (B), 283.
Cannon and Washburn. Hunger. Am. Journ. Physiol., 1912, xxix, 441.
Carlson. The Control of Hunger in Health and Disease. Chicago, 1916.
Gregersen. Am. Journ. Physiol., 1931, xcvii, 107.
Howard. Ten Weeks with Chinese Bandits. New York, 1926.
La Barre and Destrée. C. r. Soc. de Biol., 1930, ciii, 532.
Quigley and Carlson. Am. Journ. Physiol., 1931, xcvii, 107.
Quigley and Halloran. Ibid., 1932, c, 102.
Scott, Scott and Luckhardt. Ibid., 1938, cxxiii, 243.

IV

THE CONSTANCY OF THE WATER CONTENT OF THE BLOOD

I

THE importance of water in the organism has already been mentioned. Roughly it constitutes two-thirds of our weight—thus a man of average weight would have about 100 pounds of water in his composition. Naturally some parts, e.g., the bones, have less water than others. It may be surprising to learn that more than 85 per cent of the gray matter of the brain is water, more than 90 per cent of the blood plasma, whereas the saliva, lack of which, as we have seen, is the signal for need of water in the body, contains more than 98 per cent.

Water is the vehicle for food materials absorbed from the digestive canal; it is the medium in which the chemical changes take place that underlie most of our obvious activities; it is essential, as we shall see, in the regulation of body temperature; and it plays an important part in mechanical services such as the lubrication of moving parts—in the sliding of the intestinal coils on one another and the slipping of the joint surfaces to and fro.

The conservative use of water in our bodies is probably significant of its importance. There are a number of circulations of the fluid out of the body and back again, without loss. Thus the saliva leaves the body when it enters the mouth (the inside of which is, of course, not a part of the body!), in amounts varying from a quart to a quart and a half a day, and is practically all absorbed again in the intestines. The gastric juice, produced by the walls of the stomach to the extent of one or two quarts daily, and the bile from the liver, and the secretions from the pancreas and the intestinal wall, amounting perhaps to two quarts—each composed almost wholly of water—are discharged to perform their special functions of delivering the active ferments for the processes of digestion. Then they pass back into the body through the inner lining of the intestinal canal, carrying as they do so the digested food. A somewhat similar circulation of watery fluid is seen in the kidneys; water and substances in solution go freely through the little cup of capillary vessels at the top of each one of the myriads of minute tubules which constitute the essential portion of the kidney, but in passing through the tubules some of the water, the useful salts and the dissolved sugar are retrieved, leaving the useless waste to flow away. In these various circulations, as we can readily understand, water participates as a vehicle in very important transportations, without, however, departing permanently from the body.

Further evidence of the value of water to the organism is furnished indirectly by comparing the effects of its loss with the loss of other materials. According to the German physiologist, Rubner, we can, by fasting, lose practically all of our stored animal starch or glycogen, without noteworthy consequences; we can likewise lose all our reserves of fat and about one-half the protein which is either stored or built into body structure, and not be confronted by great danger. On the other hand, a loss of 10 per cent of the body water is serious and a loss of 20–22 per cent means certain death.

II

Quite possibly the grave effects of water loss are due to the changes induced thereby in the composition of the common carrier, the blood. No great variations can occur in the water content of the blood without producing marked disturbances. In dysentery and cholera, for example, the diarrhea carries water away continuously from the body, and does not permit restorative fresh water to be absorbed when it is drunk. In consequence, the blood volume becomes reduced by loss of some of its water content; and the composition of the blood is thickened, as indicated by concentration of the red corpuscles and the plasma and by the increased weight of a given volume. In association with these changes the blood becomes a more and more viscous fluid, having so much internal friction that only with difficulty can it be kept circulating. The corpuscles stick in the capillaries and fail to return in normal amount to the heart, the heart is unable to discharge its normal output, the blood pressure therefore falls, and a shock-like state supervenes. Some time before any such disastrous extremity is approached, however, mere reduction of the water content of the blood, as Woodyatt's experiments have shown, may bring about a state of fever.

There may be hazard, also, if too much water is present

in the blood. By taking the drug, pituitrin, it is possible to prevent the loss of water through the kidneys. When, in these circumstances, a large amount of water is drunk, a condition of "water intoxication" is induced, characterized by headache, nausea, dizziness, weakness and incoördinated movements. This state is rather artificial, and is not nearly so likely to be disturbing as the concentration of the blood by water loss.

III

We are continuously and necessarily losing water from our bodies. In breathing, as we have noted, we take in oxygen and give off carbon dioxide. These gases can pass rapidly through the walls of the little air sacs of the lungs only if the walls are kept moist. Not only that part of the respiratory tract, but the upper parts, in the nose and throat, in the trachea and its bronchial branches, are all kept lined with a coating of fluid. Unless the air breathed in is thoroughly laden with moisture, each breath that goes out carries away water from the body, as every child has demonstrated by breathing on a cold window pane. On a dry day it is estimated that the amount of water which is thus carried away may amount to nearly a pint.

Water is also lost from the body as perspiration. This may be "insensible," i.e., imperceptible by common tests, or it may be outstanding sweat. For a man at rest or engaged in minor activities the average loss of water by insensible perspiration has been estimated by Dr. and Mrs. Benedict as being about the same amount that is lost by the breath—nearly a pint. Perspiration is insensible when bodily activity is slight and the surrounding air is not very humid; the small amount of sweat is then so quickly vaporized that it is not noticed. When the weather is hot, or the clothing is too warm, or extra heat has been produced by muscular work, the secretion of sweat is increased. It may be greatly increased. In a football game or in a strenuous race a man may lose 4 or 5 pounds, mostly water. A large part of this loss is through the skin.

A third way in which water is constantly leaving the body is by the kidneys. We must remember that by that route are discharged from the body the non-volatile waste products. They are continuously resulting from activity. They must be removed from the blood in order to keep its composition uniform. They can only be removed in solution in water. To get rid of 45 grams of the waste substance, urea, requires about a quart of water. Marriot has described a patient, living without food or water, whose non-volatile waste required the loss of nearly a pint of water daily through the kidneys, i.e., the patient had an unreplaced loss of that amount by that one channel.

From all these observations it is clear that in the face of the dangers which arise when blood loses water there is a necessary and large drainage of water away from the body, going on without interruption.

IV

I have previously called attention to the remarkable fact that deprivation of water for days may be accompanied by no change in the blood. Wettendorff, in Brussels, examined the blood of dogs who had gone without drinking for three days, and with the refined test which he used he observed no alteration in its consistency. After four days there was a barely perceptible change. Although the dog has no sweat glands, he loses much water, as man does, by the lungs and the kidneys; but despite this outgo and no intake, in the experiments of Wettendorff and also those of André Mayer, the water of the blood was kept constant.

A constant state is also maintained in the blood when large amounts of fluid are taken into the body. This fact was strikingly demonstrated in experiments performed on themselves by Haldane and Priestley. They reported the extraordinary feat of drinking 5.5 liters of water (nearly six quarts) in six hours. The rate of output through the kidneys rose at one time as high as 1200 cubic centimeters (about 1¼ quarts) per hour. The volume of water borne by the circulating blood from the intestines, where the water was absorbed, to the kidneys, where it was discharged, exceeded by one-third the total estimated volume of the blood! And yet in tests made on the color of the blood during the period no appreciable dilution of the blood could be observed.

Whether the intake of water is greatly reduced or greatly increased, therefore, the rapidly flowing part of the fluid matrix is kept in a remarkably uniform condition. We must next inquire into the devices for managing this steady state.

V

According to our present knowledge the regulation for ridding the organism of unneeded water is performed directly by the kidneys. As Cushny remarked, however, it is a "regulation for excess, and not for deficiency." Even when a person is suffering from a dearth of water he is forced to lose some water through the kidneys, as we have noted, in order that non-volatile waste may be carried away. The kidneys display their regulatory functions whenever the intake of water or watery fluid has been abundant. In such circumstances the kidneys reveal not only an astonishing capacity to remove large amounts of water in a given time, but also an extremely fine sensitiveness to very slight alterations in the composition of the blood. Although in the experiments of Haldane and Priestley measurements of the color of the blood showed no sign of dilution, later experiments by Priestley, who employed an electrical method for testing, proved that when a large volume of water is drunk, the electrical conductivity of the blood is perceptibly decreased. Also the osmotic pressure is slightly but demonstrably less. The kidney apparatus must be acutely responsive to such an apparently negligible change, and with admirable efficiency prevents it from becoming large.

I have referred already to the tuft of capillaries lying in the microscopic cup at the top of each one of the millions of minute tubules which make up the bulk of the kidney. Water (with urea, salts and sugar) is filtered out from the capillaries into the tubule at the cup (or glomerulus). The albuminous portion of the blood plasma does not pass the filter, however, and, just as in the relation of the plasma and the lymph, the difference in the albuminous (or protein) content of the plasma and of the glomerular filtrate results in an osmotic pressure. This works in the direction of lessening the force of the filtration pressure. Here we have illustrated the interaction of constants in the fluid matrix, for it is clear that only by having a uniform concentration of protein in the plasma will there be a uniform limitation on the filtering process, i.e., on the passage of water out of the blood. In this sense, therefore, the constancy of the water content of the blood depends to a high degree on the constancy of the plasma proteins.

On the basis of present ideas the water and the substances dissolved in it, after filtering through the glomeruli, are markedly affected by the treatment they undergo in passing down the tubules. The cells which line the tubules are of such form as to indicate an ability to work, and there is evidence that they actually do work. According to a widely accepted theory, their efforts are spent in absorbing back into the body again, from the glomerular filtrate, a combination of water, salt and sugar such as is normally found in the blood. That leaves behind, as escaping waste, urea, uric acid, and other acid substances, dissolved in such water as is not absorbed or as may be requisite as a vehicle. If great excess of sugar or salt has been taken in the food, it may be present in the fluid of the tubules in such large amount as to oppose by osmotic pressure the absorption of water by the cells in the tubular wall. We can understand, therefore, that regulation of constancy of salt and sugar in the blood, as well as constancy of the plasma proteins, determines largely the water content of the body.

VI

We have now seen how the consistency of the blood is kept uniform when water is taken into the body in excess.

How is it kept uniform when for a long period no water is taken in? The evidence is quite clear that water is stored and that it is released as it is needed. Engels' experiments demonstrated that after physiological salt solution was slowly introduced into a vein for an hour, nearly 60 per cent was retained. Examination of different parts of the body at the end of the injection brought out the fact that most of the fluid was in the muscles and the skin. It is interesting to observe that the blood itself was only a little changed although more than a quart of the salt solution (1200 cc.) had entered it.

The evidence that water is stored chiefly in the muscles and skin is further fortified by a study of the various organs after hemorrhage. We have already learned that bleeding is followed by passage of water from lymph to blood. In this process all tissues lose water. By comparing the organs of one side of the body with those of the other side in the same animal (the cat), before and after a considerable loss of blood, Skelton found that most of the water derived from the tissues after the bleeding comes from the muscles and the skin—about 14.5 per cent from muscles and 11 per cent from the skin, in animals well supplied with water, and 16 and 43 per cent, respectively, in thirsting animals. These observations indicate that the muscles are the main seat of the reserves, but muscles constitute nearly half of the bulk of the body, and in reality they lose less water per unit of weight than do other parts.

The entrance of water into these storage places appears to be a sort of *inundation*. I have already likened the lymph spaces to a swamp in which fluid stagnates. That analogy is implied also by the word "inundation." We may think of the tissue spaces as being a sort of bog into which water soaks when the supply is bountiful and from which the water seeps back into the distributing system (the blood vessels) when the supply is meager. There seems to be an arrangement for that sort of relation in the loose, fine-meshed connective tissue found especially under the skin and around and between muscles and muscle bundles, but also in other parts of the body. Connective tissue is distinguished from other kinds in being richest in extracellular colloid materials, in having a close association with blood vessels—indeed, it serves as a support for blood vessels—and in exposing an enormous surface area. In such structures, chiefly, do the agencies rule which hold not only mobile water, but also substances dissolved in it, such as salts and sugar (glucose). Here few cells are found, but instead "a spongy cobweb of delicate filaments" which are held together by a small amount of "cement substance." Within the tiny interstices of this microscopic net of collagenous fibers are found protein substances (mucoid and small amounts of albumin and globulin). In this mesh, and bound by it in some manner, water and its dissolved substances appear to be stored. And it is in this mesh that the fluid collects when the heart or the kidneys fail to perform properly their functions. Edema or "dropsy" results, with swelling about the ankles or puffiness under the skin here and there, because the accumulation in the connective tissue is no longer normal.

Although the evidence just presented points to alveolar connective tissue as the reservoir for the water supply of the organism, there is a possibility that in case of dire need the water within certain cells, the cell sap, may be taken from these cells for the benefit of other cells. We have already seen that when the blood pressure falls after hemorrhage the volume flow of blood to peripheral organs is reduced, to the advantage of organs essential for continued existence. Similarly in starvation, some tissues waste away, i.e., yield some of their structure, but not the tissues of the heart and brain. These have preferential treatment, and at the end of a prolonged fast are found quite normal. It may be that the heart and brain are well provided with water even to the end of a disastrous term of thirsting, at the expense of other organs.

VII

After hemorrhage or profuse sweating water is brought out from storage abruptly and rather rapidly, at a time when there is critical need for it. But apparently it is passing out continuously from these storage places to meet the requirement which is always presented of preserving the homeostasis of the blood in the face of continuous losses through the lungs, the sweat glands and the kidneys. Only by assuming that there is an arrangement whereby water is liberated from the reservoirs as needed can we explain the striking results observed by Mayer and by Wettendorff in thirsting dogs—the unchanged state of the blood after several days of water deprivation.

Just how water is released from storage as it is needed to keep constant the state of the blood is not yet fully explained—indeed, the same statement may be made regarding its deposit in the reservoirs and its retention there. We know that the sap of the cells, like the blood plasma and the lymph, is a watery solution of salts, sugar and albuminous materials. Between this sap and the lymph everywhere is the membrane of the cell wall through which water and certain of the dissolved substances can readily pass. Ordinarily the water of the plasma is balanced against that in the lymph, and that in turn is balanced against the water in the cell sap. If the water of the plasma is increased it disturbs the balance, and we may suppose that it diffuses outward into the lymph. On the other hand, if the water of the plasma is decreased and the blood becomes thickened, we may reasonably assume that water comes out from the lymph to the blood. It was with such an arrangement in mind that I suggested earlier the idea of "inundation" into the tissue spaces when the water intake is abundant—a process which, naturally, is limited by the balances just mentioned and also by the overflow through the kidneys—and the seepage back into the plasma when the water outgo from the body tends to alter the consistency of the blood. But this may not be the whole story.

There is evidence, as Adolph, and Baird and Haldane have found, that the taking of common salt (sodium chloride) with water will markedly increase the retention of water in the body. There is evidence also that slight shifts in the direction of an acid or alkaline reaction will affect the storage. According to observations made by Schade, a shift towards an alkaline reaction causes a retention of water by connective tissue, and a shift in the opposite direction, i.e., towards the acid side of neutrality, results in release. It may be, also, that the thyroid gland, in the neck, plays a rôle through its internal secretion. When this gland is diseased or removed there is a great accumulation of albuminous material and water in the connective tissue under the skin—a condition known as myxedema. The condition can be readily cured by treatment with thyroid extract or with the essential part of the extract, thyroxin. When that is done the disappearance of the myxedema is associated with a large release of water and salt which pass out through the kidneys.

As I have already said, it is not clear how these various factors coöperate when water and salt are needed in the circulation, as, for example, after an abrupt hemorrhage. The evidence that we have reviewed in the foregoing pages, however, proves that the storage and the release of water are of primary significance for the organism. Our ignorance of the part played by various factors in these processes measures the need for further investigation.

VIII

It is admitted that we are not completely informed regarding the details of the transfer of water to and fro between the blood and the tissue spaces (and possibly also the cells). We do know, however, that the water is stored. And we know that it comes forth from storage at such a rate as to keep the blood uniform, although the blood is subjected to a continuous drain. In other words, the motile part of the fluid matrix is assured constancy to a degree that is amazing, when we consider the various disturbing conditions, always present, which may alter it. If water is drunk in a large amount, it is not allowed to dilute the blood, but is either reserved in the connective tissue spaces or discharged via the kidneys. And the losses of water from the body by sweating, by breathing, by the secretion of urine, or by temporary escape into the alimentary canal, likewise do not alter the composition of the blood to any noteworthy degree. In these circumstances it is kept uniform by contributions from reserves in the tissues. As we have noted, the largest amounts are surrendered by the skin and the muscles. Other parts, as well, give up their holdings, or suffer from an inadequate supply because of the scarcity. Among these other organs are the salivary glands. The saliva which they produce is, we have learned, more than 98 per cent water. When their water supply runs low, therefore, they can not produce an adequate amount of thin saliva to keep the mouth and throat comfortable. The disagreeable sensations of dryness and stickiness which arise from that region we recognize as thirst. Thirst leads to the drinking of water or watery fluids and thus to the restoration of the normal reserves in the body—and to resumption of the normal service of the salivary glands.

References

Adolph. Am. Journ. Physiol., 1923, lxv, 419.
Baird and Haldane. Journ. Physiol., 1922, lvi, 259.
Benedict, F. G. and C. G. Proc. Nat'l Acad. Sci., 1927, vi, 364.
Cushny. The Secretion of Urine. London, 1926.
Engels. Arch. exp. Pathol. u. Pharmakol., 1904, li, 355.
Haldane and Priestley. Journ. Physiol., 1915, 1, 296.
Mayer. C. r. Soc. de Biol., 1900, lii, 154, 389, 522.
Schade. Oppenheimer's Hdbuch. d. Biochem., Jena, 1923, viii, 172,

V

THE CONSTANCY OF THE SALT CONTENT OF THE BLOOD

I

IN ORDER to avoid complications the regulation of water in the organism was considered in the last chapter with only few references to a corresponding regulation of some of the salts dissolved in it. In the plasma and lymph are found sodium chloride (NaCl), potassium chloride (KCl) and also calcium chloride ($CaCl_2$), as well as phosphates and sulphates of the three bases (Na, K and Ca). By far the largest constituent of the mineral matter of the blood is common table salt, NaCl. Although all these salts are important for the proper functioning of the body, we shall not be able to study the regulation of all of them. We shall examine the way in which sodium chloride is kept constant in the blood, because of its close relation to the control of the water level, and we shall later take up the government of the calcium concentration, which is managed quite differently.

II

There is good evidence that the sodium and the chloride parts (the sodium and the chloride *ions*) of the NaCl in the plasma may vary independently, and that of the two the base is much the more constant element. In a study of the steady conditions in the fluid matrix, therefore, the emphasis might properly be laid on the homeostasis of the basic ions. Since most of the facts now available, however, have come from experiments in which the behavior of sodium chloride, as such, was investigated, I shall consider the substance as a salt.

The relatively large amount of sodium chloride in the plasma and lymph makes it an important agent in the osmotic relations of these fluids. If its percentage in the plasma rises—for example, from 0.56 to 1.0 per cent—the osmotic properties are much altered. Increase of osmotic pressure in the body fluids causes withdrawal of water from the cells. Constancy of the concentration of salt in the blood is evidently important.

If a diet lacking sodium chloride is fed to an animal for a sufficiently long time the amount of the salt in the blood may be considerably reduced. The kidneys do not let more than a trace of it escape. If now the protective check on the outgo is removed, by giving a diuretic drug which causes the salty water to be flushed through the kidney tubules, much salt is excreted again. This excessive loss brings on an increased irritability, then weakness and shivering, and finally paralysis of the hind limbs and a few hours thereafter death. If a solution of sodium chloride is injected, the animal is restored to the normal state, unless the effects of the deprivation are too extreme. These observations, which were made on rabbits by Grünewald, demonstrate the importance of keeping up the salt content of the blood to the standard level. Taylor has reported the symptoms which he observed in himself while living on a salt-free diet. Sweating became prominent, he early lost appetite, on the fifth day he experienced noteworthy lassitude, on the eighth and ninth days he suffered from muscular soreness and stiffness, and then from sleeplessness and muscular twitchings. Further indications of even more abnormal developments led to the stopping of the experiment.

If salt is given in excess to children it may bring on a fever, called "salt fever." The phenomenon can readily be produced in lower animals by intravenous injections of rather concentrated salt solution. Ordinarily, however, a considerable excess of salt, above the actual needs, may be taken into the body without causing any noteworthy disturbance.

The foregoing data prove the ill effects of a reduced content of NaCl in the blood and the ability of our bodies to receive and dispose of an indefinitely large amount of it. This situation is notably similar to that of water in the organism. Like water, salt is lost from the body continuously, both in the urine and in sweat. The constancy of the concentration in the blood suggests that it, like water, is stored somewhere in the body, and released as it is needed.

III

That salt may be retained in the body was shown by Baird and Haldane in experiments on themselves. They swallowed concentrated solutions of sodium chloride and sodium bicarbonate ($NaHCO_3$), and found that only a part of the amount ingested appeared in the urine. Most of it was held back in the body. Now the interesting fact appeared that drinking rather large amounts of water (two quarts or more) did not wash the salt out of its place of deposit in the tissues. It would seem, therefore, that salt, when stored in the tissues, is given up by them somewhat slowly.

Further evidence of storage was offered by Cohnheim and Kreglinger. In climbing Monte Rosa one of them lost more than six pounds, mostly in sweat. The daily intake of salt had been fairly uniform. The output through the kidneys, however, on the day of the climb, was greatly reduced —a fact to be associated, no doubt, with the large loss of salt in the sweat. But there was also a low output through the kidneys *the next day*, when the experimenter was resting. Comparison of the amount of salt taken in and the amount discharged revealed that, after the heavy losses on the day of the climb, the body retained between 10 and 14 grams of salt, i.e., nearly four teaspoonfuls. These observations suggest that the bodily reserves of NaCl were somewhat depleted by the sweating and were then filled up again from the salt that was eaten on the day of rest.

IV

If salt is stored in the body, where is the storage place? Some studies made by Padtberg have shown that the lungs, the kidneys, the blood and the skin have the highest percentage of sodium chloride. With a salt-rich diet, one-third of the salt of the body may be in the skin; and after an intravenous infusion of salt solution, between 28 and 77 per cent of the retained salt is found in the subcutaneous connective tissue. Padtberg discovered also that after a diet poor in salt had been fed for some time the chloride content of the body may fall between 11 and 21 per cent, and that between 60 and 90 per cent of this amount is accounted for by losses from the skin layer. Only to a meager extent do other organs share the function of storing salt. Of course, in the skin it is stored in combination with water and the other substances in the minute reticulum of connective tissue strands which give the spongy character to the region. And when water is withdrawn from this reservoir, we must suppose that salt is withdrawn with it, for otherwise the osmotic pressure of the plasma would be altered.

V

Undoubtedly the fine adjustment for the control of the constancy of the salt content of the blood is located in the kidney tubules. The uniform level of the base, sodium, in the blood is 0.3 per cent. If the percentage rises above 0.3 it passes through the cup-like glomerulus with water and, according to current ideas, only the combination of salt and water that corresponds to the normal ratio in the blood is taken back by the cells along the lower reaches of the tubules. Thus the excess of sodium is disposed of. In the process the concentration in the urine may rise as high as 2 per cent. If, on the other hand, the sodium content of the blood tends to fall, the salt is arrested in the tubules and returned to the body. Water and salt still filter through the glomerulus, but as usual the combination of salt and water conforming to their normal ratio in the blood is reabsorbed, and since the salt is not present in excess it is practically all saved. In the experiments which Taylor performed on himself, when he lived for more than a week on a salt-free diet, the total output of chlorides daily was reduced to the low figure of 0.2 gram. The sweat glands are somewhat like the glomerulus, but without the absorbing tubule attached. Salt passing out with the sweat, therefore, is permanently lost from the body.

When no salt is ingested for some time, we may, on the basis of the facts at hand, assume that the salt content of the blood is kept uniform by a withdrawal of salt from storage in the skin, and at the same time by a protection against loss, so far as possible, through the conservative action of the kidney tubules.

If there is need for salt in the body, as there may be, for example, in herbivorous animals whose diet contains more potassium than the body requires, the phenomenon of "salt hunger" appears. Authentic reports are on record that such animals travel long distances to "salt licks" to satisfy their "hunger." The nature of this hunger is quite unknown.

The type of homeostatic arrangement which keeps constant the salt level in the blood appears to be the same as that of water. We may assume an inundation into the boggy mesh of the connective tissue for storage, and an overflow through the kidneys when the supply is excessive. And if a salt deficit arises, it is compensated for by an outflow from the storage stations, and at the same time a reduction of the wastage in the urine. "Salt hunger" may be analogous to thirst as a means of meeting the gross requirements of the organism. In practically all respects the control of water and the control of salt are alike, and it is probable that they usually run parallel.

Evidence for parallel changes in the salt and water content of the fluid matrix is found in the effects of removal of the adrenal cortical tissue. Because removal of the adrenal medulla does not produce the typical changes, they may be attributed to lack of the cortex when the glands are extirpated intact. After dogs have been deprived of the adrenal glands for a day or two the blood begins to show a progressive reduction of the sodium-chloride content and also the plasma volume. The count of red corpuscles rises reciprocally and, for some reason not yet clear, the potassium concentration of the blood does likewise. Since these disturbed conditions can be reversed and corrected by administration of cortical extract (cortin), as shown by Zwemer and Sullivan, it appears that among the functions of the adrenal cortex is that of maintaining homeostasis of water and sodium chloride, and possibly potassium also, in the circulating blood.

References

Baird and Haldane. Journ. Physiol., 1922, lvi, 259.

Cohnheim and Kreglinger. Ztschr. f. physiol. Chem., 1909, lxiii, 429.

Grünewald. Arch. f. exper. Pathol. u. Pharmakol., 1909, lx, 360.

Padtberg. Arch. f. exper. Pathol. u. Pharmakol., 1910, lxiii, 78.

Taylor. Univ. of California Publications, 1904.

Zwemer and Sullivan. Endocrinol., 1934, xviii, 97.

VI

THE HOMEOSTASIS OF BLOOD SUGAR

I

GRAPE sugar or glucose is the form into which starchy food is changed in order to be suitable for use in the body. Of all energy-yielding materials supplied by the food, glucose is the most readily serviceable. When it is provided in abundance it is preferably utilized; the burning of fat is then almost completely stopped. Furthermore, according to present views, glucose or its storage precursor, glycogen, is essential for muscular contraction. The substance is continuously being used, therefore; even during sleep the heart muscle and the muscles of respiration are consuming glycogen, and it can be renewed only periodically.

Ordinarily the concentration of circulating glucose is 100 milligrams in 100 cubic centimeters of blood—commonly expressed as "100 milligrams per cent." It should not vary greatly above or below that concentration. If after a meal rich in sugary food, or after the eating of a large amount of candy, the concentration rises above the "kidney threshold" (about 180 milligrams per cent), the sugar is lost from the body in the urine. If on the contrary the concentration falls to 70 milligrams per cent or lower, the "hypoglycemic reaction" is likely to appear.

The hypoglycemic reaction has been given prominence by the use of a preparation of the internal secretion of the pancreas, insulin, which is employed in the treatment of diabetes. The rationale of its use in that disease we shall consider later. At present we may merely note that its action in producing a fall of the blood sugar may go too far. When, under the influence of insulin, the glycemic (blood sugar) level is reduced to about 70 milligrams per cent, the patient usually complains of a sense of weakness or fatigue and experiences the pangs of hunger. Almost always there is a feeling of tremulousness, and some incoördination for fine movements. If the sugar concentration is further reduced, objective signs appear; sweating is very profuse, pallor and flushing are common, the pupils dilate, and the pulse rate is accelerated (especially in children). At the same time the subjective symptoms become more severe; the nervousness turns to anxiety, to excitement and even to an emotional outburst. If the fall of the blood sugar is not checked, alarming manifestations may present themselves, such as great emotional instability, disorders of speech, mental confusion and delirium.

Similar phenomena, reaching a climax in convulsions and coma when a level of about 45 milligrams per cent is reached, are displayed by lower animals which have been treated with too much insulin. That the effects are not due to insulin itself, but result from the reduction of the sugar content of the blood by insulin, was proved by the experiments of Mann and Magath. They found that when the liver was excluded from service to the body, the glycemic level fell and that convulsive movements, followed by unconsciousness, were characteristic of a hypoglycemia of 45–50 milligrams. That is, the results were the same as those produced by insulin hypoglycemia. A marvelous transformation is produced both in lower animals and in human beings, who are suffering from hypoglycemia, if glucose is injected into the blood stream; the alarming symptoms or the indications of an apparently moribund condition miraculously vanish and almost immediately the normal status is recovered.

Sugar, then, is continuously being used in the body. It is only periodically renewed. The delivery to the blood must not only be continuous, but must be so adjusted to the demand that there is not an oversupply which results in loss of valuable energy-yielding material from the body and also not an undersupply which may cause a more or less profound disturbance of the whole organism.

II

Homeostasis of sugar in the blood is secured by storage as an intermediary between abundance and need. But the storage of this material differs from that of water and salt in occurring in two stages.

The first, temporary depository for excessive blood sugar, as for excessive sodium chloride, is in the skin. When sugar or other readily digestible carbohydrate food is a large constituent of the diet, the glycemic concentration commonly rises from about 100 to about 170 milligrams per cent—i.e., to a point just below the kidney threshold. Folin, Trimble and Newman have found that during this period, when the percentage of sugar in the blood is high, the percentage in the skin is also high. Again this appears to be an example of storage by inundation. No chemical change occurs in the glucose. No special device is required either to deposit it in the temporary reservoir or to remove it therefrom. As the circulating sugar is utilized or given more permanent storage, in a manner which we shall soon consider, the level in the blood falls. Thereupon the more concentrated glucose, which has overflowed into the spongy interstices of the deep layers of the skin, and possibly into other regions where alveolar tissue abounds, gradually diffuses back into the blood again, and then follows the usual courses of the blood glucose into immediate use or into the more fixed reserves.

The second stage or mode of storage, which is typical not only of the economy of carbohydrate but of other materials as well, is that of inclusion in cells or special places. This may be called storage by *segregation.* It differs from storage by inundation in being subject to much more complicated control. Storage by inundation, as we have learned, may be conceived as a process of outflow from the blood stream and backflow into it according to the degree of abundance—a relatively simple process. Storage by segregation, on the other hand, usually involves changes of physical state or of molecular configuration, and appears to be subject to control by the nervous system or by that system in coöperation with glands of internal secretion. This statement is necessarily tentative because of the large gaps in our knowledge, which further consideration will reveal.

Carbohydrate is stored in plants in the form of starch. It

is stored in animals likewise in the form of "animal starch" or glycogen. The circulating form in both plants and animals is sugar in solution in a watery fluid. Witness the maple syrup from the spring flow of sap in the hard maple. The glucose of the flowing blood is changed to glycogen in reserves which are set aside in the cells of the liver and of muscles. When required for use it is changed back to glucose again by the liver cells and then may be carried in the blood to regions where it is needed. The glycogen of the muscle cells is changed to lactic acid; this also may enter the blood and on reaching the liver, it is interesting to note, may there be built back into glycogen again.

The arrangements for storage and release of carbohydrate afford the best example of homeostasis by means of segregation. When carbohydrate food is plentiful, the glycogen reserves in the liver are large. After prolonged muscular work, however, these reserves may be almost wholly discharged. It is important to observe that while the discharge is going on it is evidently under control. Campos, Lundin, Walker and I have demonstrated that when dogs work vigorously in a treadmill for two hours the average level of blood sugar gradually falls from about 90 to about 66 milligrams per cent. In other words, during the period of great utilization of glucose (in muscular work) its content in the blood is maintained at concentrations which neither result in the possibility of loss through the kidneys nor threaten the possibility of serious disturbance from hypoglycemia.

III

Now let us see what happens when the blood sugar tends to increase in the internal environment. The efficacy of agencies which oppose this tendency is revealed when an excess of glucose is ingested. The blood sugar rises to a level close to that at which it escapes through the kidneys, but as a rule it does not surpass that level. The excess sugar, apart from that set aside by inundation, is either stored in the liver or in muscles, or is converted to fat, or is promptly utilized. There is evidence that the process of storage by segregation in hepatic cells and muscles is dependent on insulin. This is an internal secretion which is elaborated by groups of cells in the pancreas, the cells of the so-called "islands of Langerhans." From them it is discharged into the blood. I shall only briefly outline the evidence that in the storage process insulin plays a major rôle:

First, disease or removal of the pancreas results rapidly in the development of diabetes, with the appearance of a great excess of sugar in the blood (hyperglycemia) and a great reduction of the glycogen reserves of the liver.

Second, the administration of insulin to diabetic human beings, or to sugar-fed dogs who have diabetes, reduces the blood sugar to the normal percentage and at the same time causes glycogen to accumulate again in large amounts in the liver. In short, insulin, when injected, substitutes for what is lacking because of the deficiency of the pancreas.

Third, insulin administered to animals deprived of the pancreas causes a decided increase in the glycogen deposits in the muscles, especially when extra glucose is provided. Glucose without insulin is not thus deposited.

And finally, as evidence that the pancreas is normally involved in the control of carbohydrate utilization, there are characteristic changes which Homans has observed in the cells of the islands of Langerhans, that indicate overwork, when carbohydrate food is fed in excess to an animal having only a small remnant of the pancreas; and associated with such changes there is a functional degeneration of the cells.

Just how the pancreas is made to secrete insulin is not yet clear. There is little doubt that a high concentration of circulating glucose can stimulate directly the island cells. That is indicated by a variety of experiments. A part of the pancreas can be transplanted under the skin, as Minkowski showed, and when the rest of the gland is removed, thus breaking the nervous connections, diabetes does not appear; but when thereafter the engrafted piece is removed the disease at once becomes evident. Consistent with this result are the experiments of Gayet and Guillaumie, who showed that the excess blood sugar of experimental diabetes in a lower animal is promptly reduced when the pancreas from another animal is artificially connected with the blood vessels in the neck.

But there is evidence also that the secretion of insulin is under nervous control. After the indecisive experiments of de Corral and Macleod and his co-workers, one of my collaborators, S. W. Britton, found that, by exclusion of the sympathico-adrenal system, which, as we shall see, is opposed to the action of the pancreas, he could obtain consistent lowering of the blood sugar by stimulating the right vagus nerve. Figure 15 shows the usual course of blood sugar, after the preparatory operation under amytal anes-

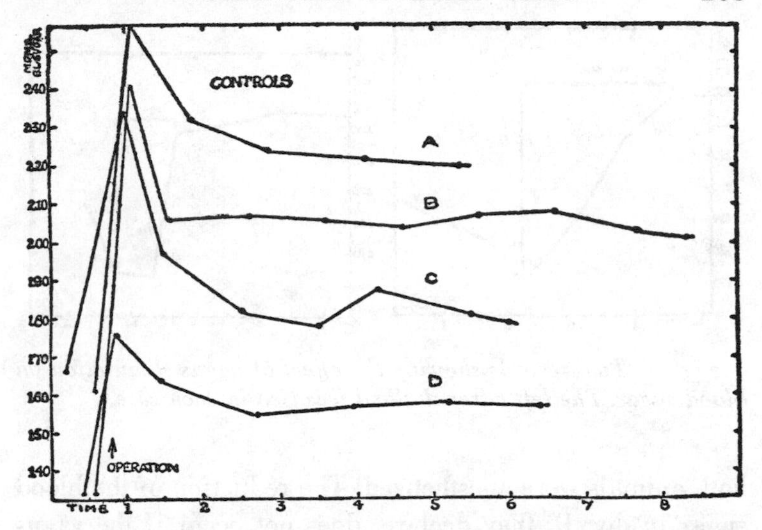

Fig. 15. A, B, C, and D. Four control experiments showing the course of blood sugar under amytal. In all cases the left adrenal was completely tied off, and the right vagus nerve was prepared as for stimulation. The operative procedures invariably produce a marked increase in blood sugar and the eventual level assumed, and maintained for several hours, is correlated with the peak of the increase.

thesia, without vagus stimulation. In figure 16 is shown the reduction of the blood sugar when in addition the right vagus was stimulated. This result did not occur if the blood vessels of the pancreas had previously been tied. According to Zunz and La Barre the nervous control of insulin secretion can be demonstrated by injection of glucose. They used a crossed circulation from the vein leading away from the *pancreas* of one dog (A) to the jugular vein in the neck of another dog (B), and found that injection of glucose into dog A caused a reduction of the blood sugar in dog B, the recipient of blood from the *pancreas* of A. Of course,

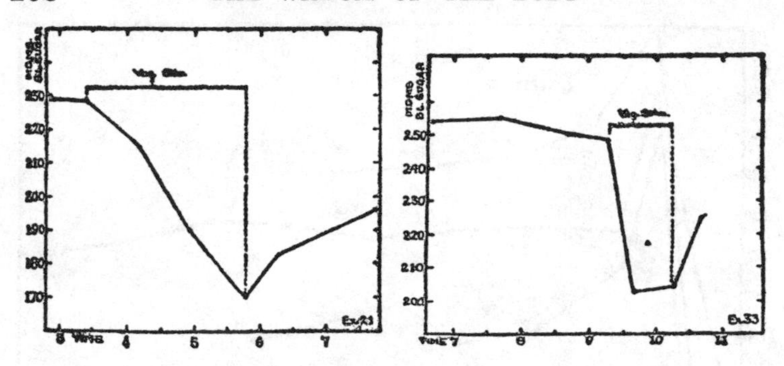

Fig. 16. Two records showing the effect of vagus stimulation on blood sugar. The left adrenal gland was tied in each case.

both animals were anesthetized. The reduction of the blood sugar in dog B, they declare, does not occur if the vagus nerves of dog A have previously been cut or if the passage of vagal impulses is blocked by the drug, atropine. Apparently the excess of blood sugar can increase the internal secretion of the pancreas via the vagi, and this augmented secretion, carried from dog A to dog B, lowered the glycemic lever in the latter animal.

The foregoing evidence, put together, indicates that a nervous control of insulin secretion exists, but that it is not necessary. That it is not necessary does not prove that it is useless. Many organs in the body can perform a sort of service though deprived of connection with the nervous system. It is possible, for example, to have a secretion of adrenin due to asphyxia even though the splanchnic nerves, which normally govern the secretion, have been cut. In these circumstances, however, the glands are not so responsive as they are when properly innervated. It may be that the vagus provides a fine adjustment for insulin secretion.

IV

Insight into the action of factors which prevent the fall of blood sugar to a seriously low level may be obtained by a study of the effects of insulin. As I have already pointed out, the reduction of the glycemic concentration to about 70 milligrams per cent by action of insulin may initiate the so-called "hypoglycemic reaction." The pallor, rapid pulse, dilated pupils, and profuse sweating, which are characteristic of the reaction, are signs of activity of the sympathetic nervous system. The question naturally arises as to whether these phenomena are part of a general display of activity by that system, and if so, whether adrenal secretion is involved. In the condition of hypoglycemia, participation by the sympathico-adrenal team would be highly interesting and significant, because that team is capable of liberating sugar from the hepatic reserves. It would be roused to activity, therefore, in a quite automatic manner by the lessened concentration of sugar in the blood, just when more sugar is needed to preserve the normal concentration.

In order to learn whether, in fact, the adrenal glands secrete adrenin in response to hypoglycemia, Bliss, McIver and I have tested the possibility by use of the "denervated" heart, i.e., the heart deprived of all its nervous connections. We have already learned that the sympathetic system sends along the fibers to the heart impulses which accelerate the beat, and also that the vagus nerves act in an opposite sense, to make the heart beat more slowly. By use of careful surgical methods Lewis, Britton and I were able to remove that part of each sympathetic chain in the upper chest from which the accelerator fibers originate, to sever the right vagus nerve below the laryngeal branch which governs the muscles of the vocal cords, and to trim away the cardiac branches of the left vagus nerve (see fig. 17). Thus the heart is quite isolated from the nervous system. Its position in the chest is not changed; it continues to pump the blood through the arteries, capillaries and veins; but its function can no longer be adjusted to the exigencies of bodily activity by direct nervous influence. The only connection of the heart with the rest of the organism is by way of the circulating blood. Fortunately for the purposes which we had in view, the rate of the denervated heart is quite unaffected by variations of arterial pressure; indeed, the only agencies, apart from temperature changes, that affect the rate are chemical substances. The denervated heart is exquisitely sensitive, for example, to a very slight increase of adrenin in the blood passing through its vessels. Anrep and Daly found that the presence of one part of adrenin in one billion four hundred million parts of blood would make the isolated heart of a cat beat faster. And Rapport and I proved that the greater the amount of adrenin brought to it in the blood stream, the greater the acceleration. The response of the heart is prompt—within ten seconds after the beginning of the discharge of adrenin from the adrenal glands the pulse accelerates.

We have used the denervated heart as an indicator of adrenal secretion in "acute experiments" (i.e., experiments in which the animal serving for the test does not recover from anesthesia), and also in experiments on animals after they had recovered from the surgical operation for severing the cardiac nerves. Cats thus operated on are soon

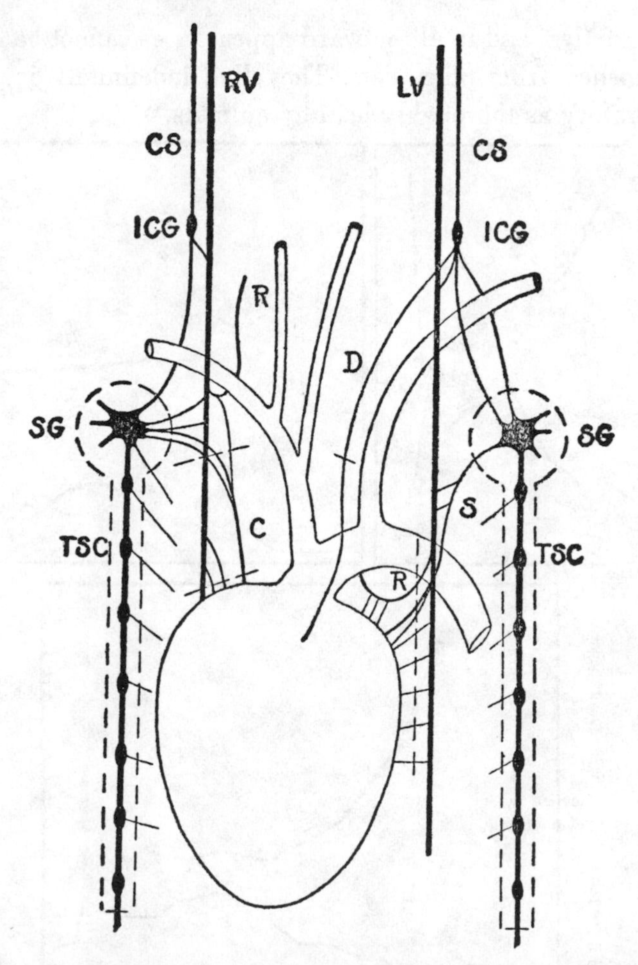

Fig. 17. Diagram of the usual arrangement of the cardiac nerves in the cat. RV, right vagus; LV, left vagus; CS, cervical sympathetic; ICG, inferior cervical ganglion; R, recurrent laryngeal nerve; D, depressor nerve; SG, stellate ganglion; C, "common cardiac nerve"; S, sympathetic fiber; TSC, thoracic sympathetic chain. The dash lines indicate the parts cut or excised.

quite active, and in all outward appearances cannot be distinguished from other cats. They live indefinitely in the laboratory as thoroughly healthy animals.

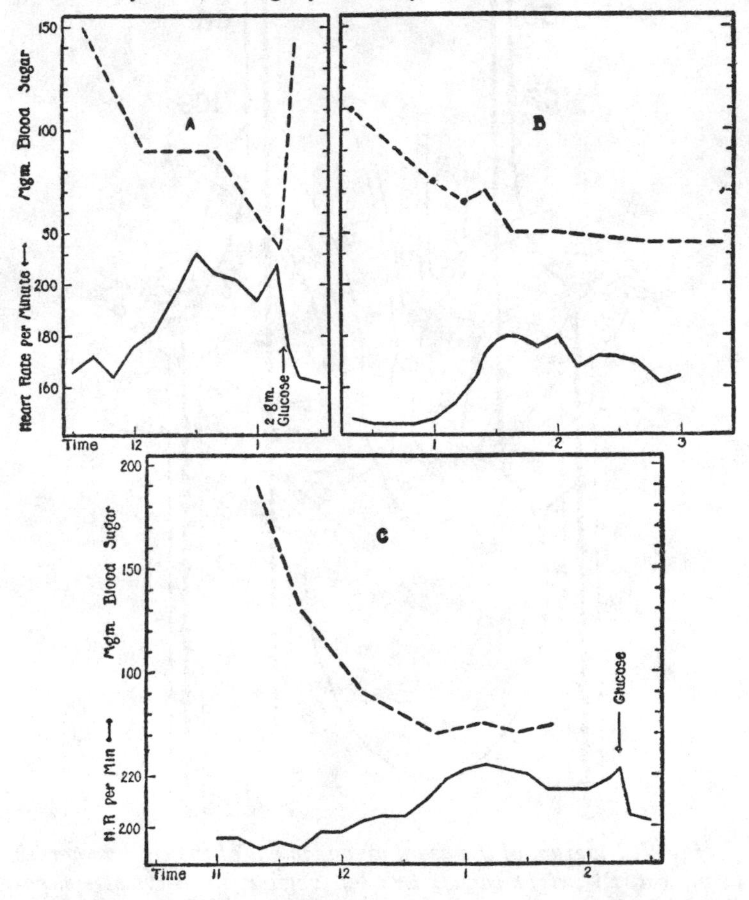

Fig. 18. Increase of the rate of the denervated heart (solid line), in animals under chloralose anesthesia, when the falling blood-sugar concentration (dash line) passed a critical point. In case A, the insulin was injected into the jugular vein at 11:33; in B, at 11:08: and in C at 9:30. In each case 4 units per kilo were injected.

By use of the denervated heart to indicate an increase of adrenin in the blood, Bliss, McIver and I found that as the blood sugar falls, after a dose of insulin, a critical point is reached at about 70 milligrams per cent in the unanesthetized animal. It is somewhat higher in the anesthetized animal. As shown in figure 18, while the glycemic concentration is falling there is no change in the heart rate until the critical point is reached. When that point is reached, however, the rate of the denervated heart begins to be accelerated and as the sugar percentage continues to fall, the heart rate continues to rise until a maximum is reached.

If the adrenal glands have previously been removed, or if one has been removed and the other denervated, a fall of the glycemic percentage, as shown in figure 19, is not accompanied by an increased heart rate. The evidence is clear, therefore, that the cardiac acceleration, which was recorded in figure 18, was not due to direct action of insulin on the heart or on the adrenal gland but resulted from an increased discharge of adrenin in response to sympathetic nerve impulses.

Now the interesting fact appears that if the rate of the denervated heart has been increased because of hypoglycemia, an intravenous injection of glucose, as shown in figure 18A, promptly lowers the rate to the former level. In other words, the sympathico-adrenal mechanism is called into action by low concentration of sugar in the blood; the effect of this mechanism would be to increase the sugar in the blood by liberating it from the glycogen reserves in the liver; if the sugar in the blood is increased by injection, and the need for sympathico-adrenal activity is thereby abolished, that activity almost immediately ceases.

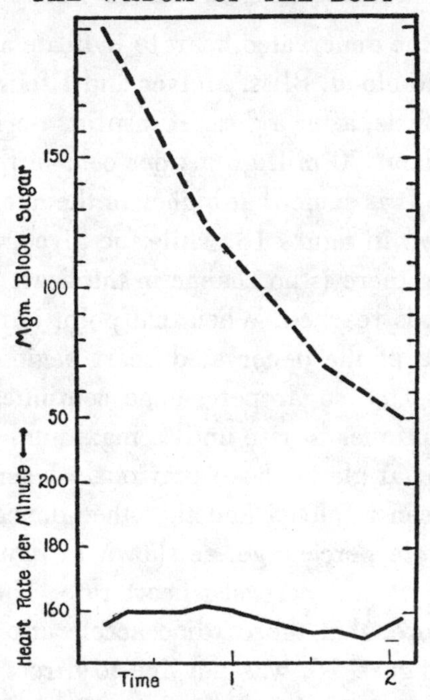

Fig. 19. Failure of increase of the rate of the denervated heart (solid line) in an animal under chloralose when the falling blood-sugar concentration (dash line) passed the critical range. The left adrenal had been removed and the right splanchnic and the hepatic nerves severed 19 days before. Insulin (4 units per k.) was injected intravenously at 12:19.

Evidence that the sympathico-adrenal mechanism does, in fact, operate to increase the sugar percentage is presented in figure 18. Observe that the rate of the decline in the glycemic percentage lessens, that is, the blood-sugar curve tends to flatten, as the sympathico-adrenal apparatus is brought into operation. Both the increased heart rate and the check in the fall of the sugar content of the blood are indicative of sympathico-adrenal action. Indeed, if not too much insulin has been given, the increased rate of the denervated heart (which, as must be remembered, signals sympathico-adrenal action) is associated with a rise of the glycemic percentage and that, in turn, is attended by a fall in the heart rate.

The value of the sympathico-adrenal system as a protection against a disturbing fall of blood sugar is shown in figure 20. Note that a series of normal cats with both adrenal glands innervated were given insulin in doses varying from two to three units per kilogram (about two pounds). In only one instance did convulsions occur, and

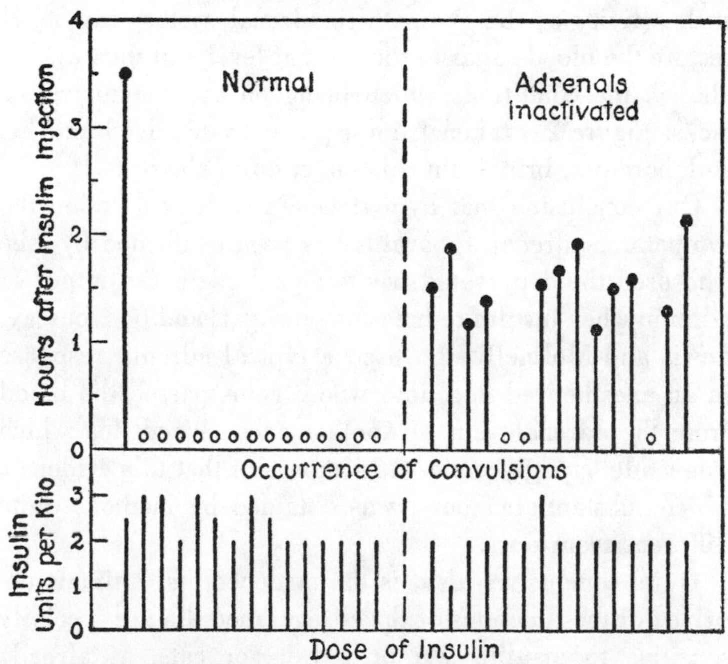

Fig. 20. Chart showing the presence or absence of convulsions after subcutaneous injections of insulin in a series of cats with normally innervated adrenals, and in a series of cats with one adrenal removed and the other denervated.

in that case they appeared three and one-half hours after the injection. Compare that series of animals with the series in which one adrenal was removed and the other denervated, that is, in which the adrenals were inactivated. Observe that, except in one instance, only two units of insulin per kilogram were employed. Yet in all but three of these cases convulsions occurred and usually about an hour and a half after the injection.

The discharge of adrenin is most active during convulsive seizures. If the liver is well supplied with glycogen such activity of the sympathico-adrenal system can itself restore the blood sugar to the normal level and thus wholly abolish the conditions which bring on the convulsive attacks. Figure 21, taken from a paper by Macleod and his collaborators, brings out this effect quite clearly.

Our conclusion that hypoglycemia calls into action the sympathico-adrenal apparatus has been confirmed by Abe who used the denervated iris to signal a greater output of adrenin when insulin reduces blood sugar, and by Houssay, Lewis and Molinelli who used a typical adrenin response in an anesthetized dog, into whose veins passed the blood from the adrenal vein of another anesthetized dog which was made hypoglycemic. It is important that this evidence, which substantiated ours, was obtained by methods quite different from ours.

Quite conclusive, also, is the evidence that animals deprived of the functions of the adrenal medulla are specially sensitive to insulin. We proved it for cats, as already shown. Lewis proved it true of rats, Sundberg proved it true of rabbits, and Hallion and Gayet observed the same phenomenon in dogs. In harmony with these are the results

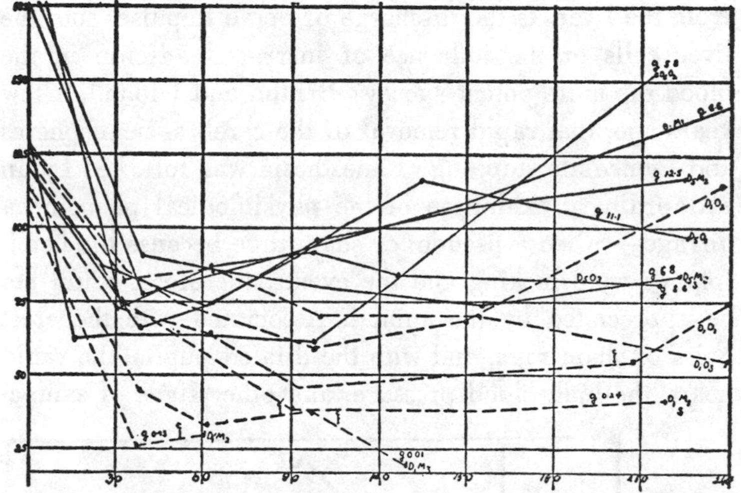

Fig. 21. Graphs showing the effect of equal doses of insulin on well-fed (continuous lines) and fasting rabbits (dotted lines). C indicates a convulsion. Note that the blood-sugar level in the well-fed rabbits sinks to approximately the critical level for the hypoglycemic reaction and then in most instances it rises. (From Macleod: "The Fuel of Life." Courtesy of The Princeton University Press.)

obtained by Burn. He used the drug, ergotamine, to paralyze specifically the sympathico-adrenal system. He discovered that a dose of insulin, which was mildly effective in a normal animal, produced in an ergotaminized animal profound hypoglycemia, with convulsions and collapse. It is clear, therefore, that it is a function of the sympathico-adrenal system to protect the organism against the deeply disturbing effects of a lowering of the glycemic level in the fluid matrix.

V

A question of some interest is the relative importance of nervous and humoral factors in the liberation of glucose from the liver. Is the discharge of nerve impulses into the liver cells or the influence of increased adrenin in the blood the more potent agency? Britton and I found, a few years ago, that rapid removal of the cerebral hemispheres and immediate stopping of anesthesia was followed by an extraordinary exhibition of the physiological phenomena of rage—called a pseudo- or sham rage because the hemispheres were lacking and the events, therefore, could not be appreciated by the animal. Associated with the erect hairs of sham rage, and with the dilated pupils, the rapid heart, the high blood pressure, and other signs of sympa-

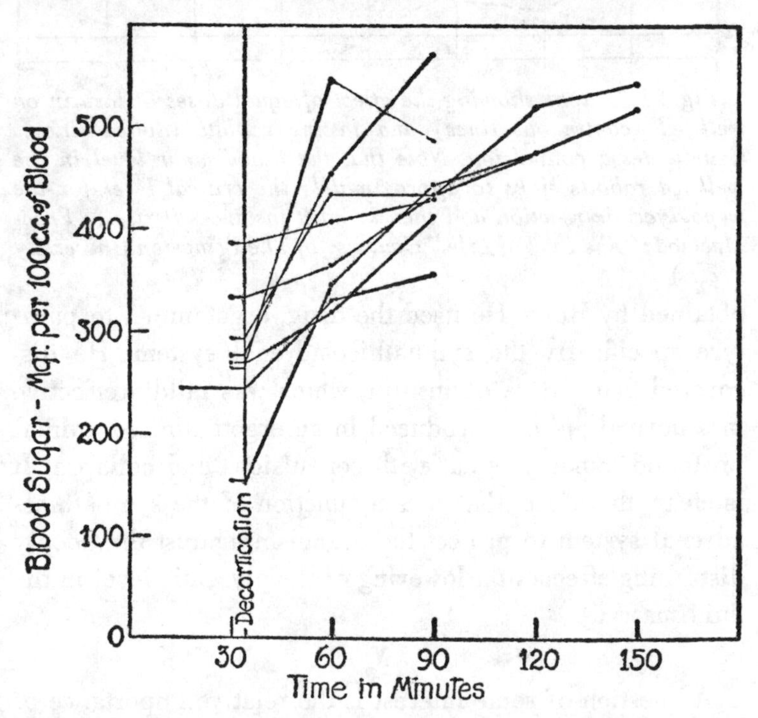

Fig. 22. Blood sugar in decorticate animals manifesting pseudaffective phenomena.

thetic innervation, Bulatao and I observed that there was an increase of blood sugar, which, as shown in figure 22, rose to five times the normal percentage. If the adrenal glands were inactivated and the nerves of the liver left, sham rage, as shown in figure 23, was not accompanied by an increase of glucose in the blood. On the other hand, if the nerves to the liver were severed and adrenal innervation remained intact, the signs of excitement were associated with nearly the same hyperglycemia as in the animal with liver nerves present. This is seen in figure 24. It

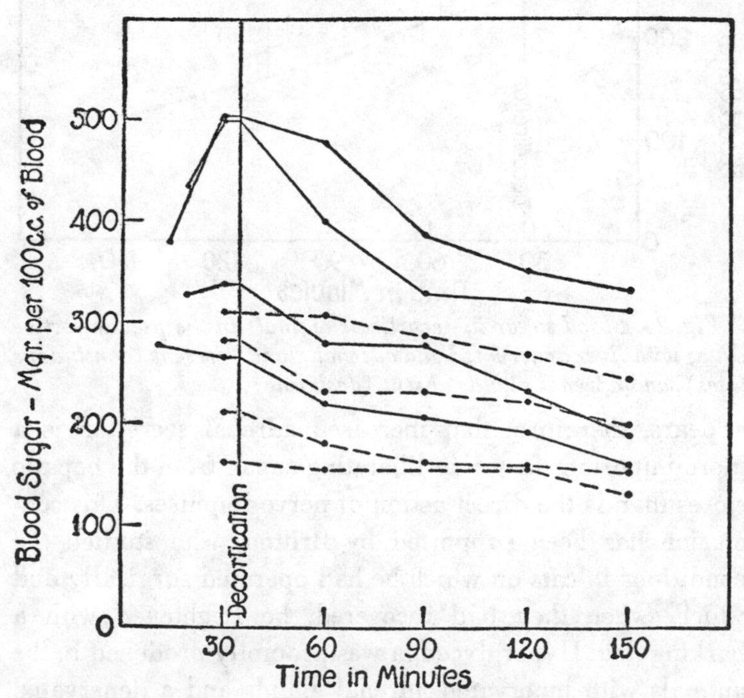

Fig. 23. Blood sugar in decorticate animals manifesting pseudaffective phenomena; without adrenal glands (continuous lines), or with the right adrenal gland removed and the left splanchnic nerves severed (dash lines).

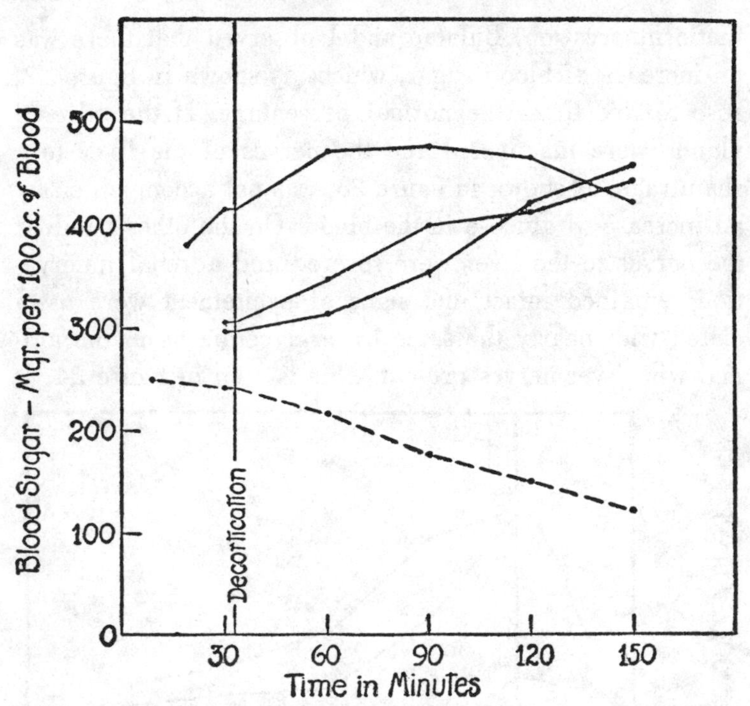

Fig. 24. Blood sugar in decorticate animals in the pseudaffective state; with liver denervated and adrenal glands present (continuous lines), and adrenal glands absent (dash line).

appears, therefore, that increased adrenal secretion is a more important factor in liberating sugar from the hepatic stores than is the direct action of nerve impulses. This conclusion has been supported by Britton, who studied the conditions in cats on which he had operated surgically and which, when they had recovered, he frightened with a barking dog. Hyperglycemia was promptly produced in the animals with innervated adrenal glands and a denervated liver, but not when the contrary condition existed.

The evidence that under natural conditions secreted

adrenin is more important than nerve impulses should not be interpreted as wholly excluding the action of nerve impulses in the breaking down of glycogen to glucose. After the adrenal glands have been removed, sensory stimulation can still cause hyperglycemia. This effect can be explained as possibly due to direct action of nerve impulses on the liver cells or to asphyxia.

VI

The general scheme which I have presented in this chapter implies the action of two antagonistic agencies which are on the alert to maintain homeostasis of blood sugar. As Hansen has pointed out, there are normal oscillations in the concentration of sugar in the blood, occurring within a relatively narrow range. Possibly these ups and downs result from action of the opposing factors, depressing or elevating the glycemic level. If known elevating agencies (normally and primarily the sympathico-adrenal apparatus) are unable to bring forth sugar from storage in the liver, the glycemic level falls from about 70 to about 45 milligrams per cent, whereupon serious symptoms (convulsions and coma) may supervene. The range between 70 and 45 milligrams per cent may be regarded as the *margin of safety*. On the other hand, if the depressing agency, the insular apparatus (i.e., the cells of the islands of Langerhans or those cells under control of the vagus nerves), is ineffective, the glycemic level rises to about 180 milligrams per cent and then sugar begins to pass into the kidney tubules in greater concentrations than can be taken back and some of it is therefore lost from the body. The range from 100 or 120 to 180 milligrams per cent may be regarded as the *margin of economy;* beyond that margin homeostasis is dependent on wasting the energy contained in the sugar and the energy which may have been expended by the body to bring it as glucose into the blood.

Besides the two agencies, insular or vago-insular and sympathico-adrenal, which act in acute conditions, there are other agencies affecting the blood-sugar concentration that seem to have a persistent and sustaining function. Thus the anterior lobe of the pituitary gland has been shown to be involved in the homeostasis of blood sugar; Houssay and Biasotti not only found that when this lobe was removed the hyperglycemia of diabetes was reduced, but also they were able to isolate from the lobe a substance which, on being injected, produced hyperglycemia. Another endocrine organ which may influence the glycemic level is the adrenal cortex; according to Britton and Silvette removal of the adrenal glands reduces both blood sugar and the stores of hepatic glycogen; and Long and Lukens have observed that the high glucose level of experimental diabetes can be lowered by adrenalectomy. These effects are not produced by taking out the adrenal medulla alone.

Although physiological factors can effect hepatic storage or release of glucose, the liver can apparently operate independently. For example, Brouha, Dill and I have noted that after removal of the sympathetic system the glycemic level is normal, even in vigorous exercise. And Soskin and his collaborators have shown precisely, by determining simultaneously the rate of blood flow through the liver and the glucose content of the inflowing and outflowing blood, that during control periods the liver secretes glucose but when glucose is abundantly supplied secretion ceases and the sugar is retained.

References

Abe. Arch. f. exper. Pathol. u. Pharmakol., 1924, ciii, 73.
Anrep and Daly. Proc. Roy. Soc., London, 1925, B xcvii, 454.
Britton. Am. Journ. Physiol., 1925, lxxiv, 291.
Britton. Ibid., 1928, lxxxvi, 340.
Britton and Silvette. Ibid., 1934, cvii, 190.
Brouha, Cannon and Dill. Journ. Physiol. (In press.)
Bulatao and Cannon. Ibid., 1925, lxxii, 295.
Burn. Journ. Physiol., 1923, lvii, 318.
Campos, Cannon, Lundin and Walker. Am. Journ. Physiol., 1929, lxxxvii, 680.
Cannon and Britton. Ibid., 1925, lxxii, 283.
Cannon, Lewis and Britton. Ibid., 1926, lxxvii, 326.
Cannon, McIver and Bliss. Ibid., 1924, lxix, 46.
Cannon and Rapport. Ibid., 1921, lviii, 308.
de Corral. Zeitschr. f. Biol., 1918, lxviii, 395.
Folin, Trimble and Newman. Journ. Biol. Chem., 1927, lxxv, 263.
Gayet and Guillaumie. C. r. Soc. de Biol., 1928, cxvii, 1613.
Hallion and Gayet. Ibid., 1925, xcii, 945.
Hansen. Acta. Med. Scand., 1923, lviii, Suppl. iv.
Homans. Journ. Med. Research, 1914, xxv, 49.
Houssay and Biasotti. C. r. Soc. de Biol., 1931, cvii, 733.
Houssay, Lewis and Molinelli. Ibid., 1924, xci, 1011.
Lewis. Ibid., 1923, lxxxix, 1118.
Long and Lukens. Science, 1934, lxxix, 569.
Mann and Magath. Arch. Int. Med., 1922, xxx, 73.
McCormick, Macleod and O'Brien. Trans. Roy. Soc. Canada, 1923, xvii, 57.
Minkowski. Arch. f. exper. Pathol. u. Pharmakol., 1908, Suppl. Bd., 399.
Soskin, Essex, Herrick and Mann. Am. Journ. Physiol., 1938, cxxiv, 558.
Sundberg. C. r. Soc. de Biol., 1923, lxxxix, 807.
Zunz and La Barre. Ibid., 1927, xcvi, 421, 708.

VII

THE HOMEOSTASIS OF BLOOD PROTEINS

I

PROTEIN food is quite as important as carbohydrate; indeed, it may be regarded as more important than carbohydrate, for it contains not only stored energy, but also certain chemical elements, including nitrogen, that are essential parts of the bodily structure. In the original upbuilding of the cellular edifice of the body, therefore, and in the repair or restoration of worn or wasted parts, and in the provision of the normal colloidal constituents of the blood, protein is necessary.

Because of stresses and motions of the bodily machinery, as it performs its daily tasks, it is continuously being subjected to frictional wear and minor injuries. The proper upkeep of the organism requires protein. There is not likely to be urgent need for large amounts of protein, as there may be for large amounts of carbohydrate when heavy and prolonged muscular work is going on, but the minute disintegrations of body structure, taking place in all its parts, are in the aggregate considerable and unavoidable.

The supply of protein for restoration and upkeep may not be continuous; among civilized men it is usually not more frequent than three meals a day, and among wild carnivora the renewal of available protein in the food may be only occasional. In the presence of a constant requirement of protein and a periodic intake, must there not be a reserve set aside for use in privation? The rapid splitting off of the part of the protein molecule which contains nitrogen, after protein food has been digested and absorbed into the body, and the prompt elimination of that part through the kidneys, would indicate that accumulation of protein reserves would certainly be limited. But even though limited, reserves might be provided.

There is evidence that protein actually is deposited in storage in the organism. Thomas lived for some time on a measured, high-protein diet. Then he stopped eating protein food and for eight days lived on a pure carbohydrate diet which was calculated to be sufficient to supply the energy for ordinary activities. During that period the nitrogen given off by the kidneys gradually fell until it was constant at 2.2 grams daily. This he regarded as measuring the unavoidable "wear and tear" of the body, i.e., the breakdown of body structure with use. At the same rate for the eight days the total amount of nitrogen given off would have been about 18 grams ($8 \times 2.2 = 17.6$). Since he actually lost 66 grams, the difference, or 48 grams (66–18), he assumed, must have been deposited as a reserve in the body. Protein is about one-sixth nitrogen; the protein deposited, therefore, would have amounted to about 300 grams (6×48), or nearly two-thirds of a pound.

Another way of calculating the account has been employed by Boothby. There is a part of the protein molecule, creatinin, which, because it is fairly constant in the urine and does not vary greatly with the taking of protein food, is regarded as measuring the disintegration necessarily going on in the cells as an accompaniment of mere existence. During a fast of thirty-one days, the professional faster, Levanzin, lost 10.7 grams of creatinin. Calculation shows that this represents 62 grams of nitrogen taken from bodily structure. But Levanzin lost during the fast 277 grams of nitrogen. It would appear, then, that 215 grams (277–62) were originally not built into the body but were in the state of deposit protein.

But more direct evidence exists that protein is stored in the organism. Results obtained by microscopic study of the cells of the liver, and chemical analysis of their contents, have agreed in supporting the conclusion that the liver can carry reserve protein as well as reserve carbohydrate. The observations by Afanassiev, in 1883, that if dogs are fed a plentiful diet of "albuminates," the liver becomes firm and resistant and the liver cells increase in size and contain protein granules between the structural strands, have been confirmed in a number of more recent investigations. These later studies have shown that when animals are given a large amount of protein food there appear in the hepatic cells fine droplets or masses. They react positively to a chemical test (Millon's reagent) for proteins; they yield, indeed, evidence of being composed of a simple protein, which disappears when the animal is made to fast, and which reappears when protein is fed.

The results of microscopic examination of liver cells conform to the results of biochemical analyses performed by Seitz and by Tichmeneff. Seitz determined the ratio of the nitrogen content of the liver to the nitrogen content of the rest of the body in fasting animals and in animals which had been similarly deprived of food but subsequently given calf meat (which is free from fat and glycogen). He found that the nitrogen in the liver, in relation to that of the rest of the body, was from two to three times as great in the fed as it was in the fasting animals. Tichmeneff's similar experiments gave similar testimony. He starved a group of mice for two days, then killed half of them. He gave the others an abundance of cooked meat and, after time for digestion and absorption, he killed them also. Then he compared the livers of the two sets of animals. Expressed in percentage of body weight, the livers of the meat-fed animals increased about 20 per cent, but the nitrogen content of the livers in these animals was increased between 53 and 78 per cent.

Confirmatory evidence has been furnished by Luck who for two weeks fed white rats with protein-poor or protein-rich diets. The protein contents of muscles, kidneys and intestines of the two groups were not markedly different. In the animals given the rich diet, however, the livers became laden with protein, as manifested by an amount of protein per unit weight of hepatic tissue much greater than in the other group.

II

We have surveyed the evidence that reserves of protein are segregated in the cells of the liver. What is the evidence that the hepatic stores have any value?

There is one part of the body, the blood, from which the protein can be removed to a measured degree and the rate and extent of its restoration then determined. Three distinct proteins are found in the blood plasma—an albumin, a globulin, and a special protein, fibrinogen, which belongs to the globulin class and which is concerned with the clot formation in blood. The albumin and globulin resemble one another in many respects but the albumin is soluble in pure water whereas the globulin requires the presence of salt to hold it in solution. The total amount of protein in the plasma is about 6 per cent, of which fibrinogen is only a small fraction (0.2–0.4 per cent).

The plasma proteins are a part of the structure of the blood and not a food substance being carried in the circulation for the nourishment of the secluded tissue cells. The reasons for that inference are found in the following facts. No regular variation in the percentage of plasma proteins is found, whether they are examined during a fasting period or during a period of heavy protein feeding. Again, if they are artificially reduced their restoration is slow, lasting several days, even though plenty of meat is eaten. And finally, restoration can occur even though no protein food is ingested.

Ordinarily the concentration of the proteins in the plasma is remarkably constant. The importance of that constancy is related mainly to the functions of these proteins as colloids. Because they exert osmotic pressure and do not readily escape through capillary walls, they prevent the salt solution of the plasma from passing freely into the perivascular spaces or out from the body through the glomeruli of the kidneys. We have already learned something of this function of the colloids of the plasma in our consideration of the factors which preserve the fluid matrix and govern its water supply (see pp. 58, 83). Some indication of the value of the plasma proteins in preserving the water and salt content of the blood is offered in experiments performed by Barcroft and Straub. They removed much of the blood from a rabbit and by means of a centrifuge separated the blood corpuscles from the plasma. They then mixed the corpuscles with a salt solution having the same volume as the plasma, and containing the same salts in the same proportions. This suspension of corpuscles they injected into a vein of the rabbit so that it was soon mixed with the remnant of the animal's blood. The only important difference then was a reduction in the colloid content of the plasma. The output of urine was soon increased to *forty times* the original rate! A part of this increase, no doubt, was occasioned by such rapid filtration through the glomerular cup that the rapid passage through the kidney tubules gave little chance for reabsorption. The experiment clearly demonstrates the eminent importance of the plasma proteins for the preservation of the water and salts of the blood.

Homeostasis of the plasma proteins, however, is not only a necessary condition for homeostasis of blood volume; one of them, fibrinogen, because of its part in blood clotting, is, in casc of hemorrhage, essential for the preservation of the blood itself. The very existence of the fluid matrix of the body is dependent, therefore, on the constancy of the proteins in the plasma.

III

We now shall inquire into the methods by which the percentage of the proteins in the plasma is kept uniform.

The information concerning these methods has come mainly from experiments performed by G. H. Whipple and his collaborators. They removed blood repeatedly from animals, separated the corpuscles, and after suspending them in physiological salt solution restored them to the circulation, in the manner already described. Thus they reduced the plasma proteins from 6 to about 2 per cent, and, in later experiments, even to 1.5 and 0.9 per cent. In all instances the percentage of proteins in the plasma promptly rose. An increase of 10 to 14 per cent occurred in the first fifteen minutes. In twenty-four hours the restoration amounted to 40 per cent of the loss. It seems quite probable that a considerable part of this recovery is relative—i.e., that because of the reduction of the colloids in the plasma, water and salt escape into lymph spaces and run away through the kidneys, and that thereby the colloids, which cannot thus depart, become concentrated. Test of a sample of the blood would thereupon erroneously indicate that the proteins had been more or less restored. Whipple has argued against that interpretation of his results. He has found that if the liver is shunted (by an "Eck fistula"), so that the blood from the alimentary canal goes, not through the liver, but directly into the inferior vena cava (see fig. 1), the rapid restoration does not occur. Also he observed that fibrinogen is not restored as quickly as the other proteins, which, of course, would not be true if simple concentration of the blood is the explanation of the recovery. And, furthermore, tests showed that the blood is not much concentrated. He draws the conclusion, therefore, that the early, sharp rise in the concentration of the plasma proteins, after it has been reduced, is due to a discharge of proteins from storage, and not to concentration because of escape of salt solution through the capillary wall. Restoration of the corpuscles, after they have been suspended in salt solution in which the albumin and globulin of the plasma are replaced by another colloid such as gum acacia, might bring out decisive further evidence on the quick rise of concentration. But, however that may be, the later, slow swing upward to the normal level is without question the consequence of some restorative process. As already stated, the increase during the first twenty-four hours brings the recovery up to 40 per cent. Thereafter the process is more gradual until the normal status is reached in two to seven days. The part played by the liver in this restoration is indicated by the effects of shunting that organ out of the portal circulation in the manner mentioned above (see p. 128); in animals with an Eck fistula, reduction of the plasma proteins may be followed by no increase at all during the first three days thereafter. And an injury to the liver, such as may be produced by the action of chloroform or phosphorus, may considerably delay the recovery.

Fibrinogen is treated differently from the other proteins. As already noted, no quick restoration takes place in the first fifteen minutes. It may, indeed, be fully discharged from storage by the hemorrhage itself. On the other hand, the normal percentage of fibrinogen in the plasma is regained by the end of twenty-four hours and probably sooner. This exceptionally rapid return does not occur if the liver is excluded from action.

IV

Although the experiments on homeostasis of plasma proteins indicate that the liver is an important source of these materials in case of need, and although the testimony cited above would justify consideration of the liver as a storage place for protein, the modes of storage and release are almost wholly unknown. Stübel did, indeed, observe that the small protein droplets or masses in the hepatic cells could be greatly reduced by injecting adrenalin subcutaneously. If these masses help to supply essential protein elements for blood clotting, as the dependence of fibrinogen on the liver would imply, their liberation by adrenalin and by conditions which would excite the sympathico-adrenal apparatus might account for certain phenomena of faster clotting. As I pointed out in Chapter II, coagulation is more rapid after adrenalin injections, after splanchnic stimulation, or after a large hemorrhage which calls the sympathetic into action, but only if the blood is allowed to flow through the liver and intestines. In this category also is the very rapid clotting of blood taken at the height of the hypoglycemic reaction when sympathico-adrenal activity is maximal.

Recently Riecker and Winters have published an account of interesting experiments on the influence of injections of adrenalin on the fibrinogen of the blood. Using the fibrin of the clot as a measure of fibrinogen, they found that subcutaneous injections of adrenalin into dogs and also into human beings caused within a few minutes a marked increase in the fibrin content of the blood. The average maximal increase was 36.3 per cent. Associated with this change was a decrease in the coagulation time, averaging about 60 per cent. Since Foster and Whipple and also Meek have brought forward evidence that fibrinogen comes from the liver, and since adrenin is known to stimulate liver cells in changing their glycogen store to glucose, it may well be true that adrenin, or the sympathico-adrenal mechanism, likewise liberates at least fibrinogen from a hepatic reserve.

It is quite possible that protein is stored in other places than the liver, and also that the thyroid gland is an important agency for controlling both storage and release. Boothby, Sandiford and Slosse have reported that with a uniform nitrogen intake a negative nitrogen balance exists (i.e., a loss of nitrogen from the body) while thyroxin, which accelerates the oxidative processes in the body, is establishing a new higher metabolic level. After its establishment there is a smaller deposit of nitrogen in the body than before. Now if the thyroid dosage is stopped (while the uniform nitrogen intake continues), a positive balance obtains until a new lower metabolic level is reached; that is, more nitrogen is deposited. These effects are much more marked in a person afflicted with myxedema than in a normal person. Indeed, as Boothby has suggested, the "edema" of myxedema may be an abnormal amount of deposit protein in and beneath the skin. The efficacy of thyroid therapy in reducing the albumin of the tissues in cases of myxedema, supports the view that the thyroid gland is somehow associated with protein regulation and metabolism.

Although the foregoing considerations have emphasized the primary importance of homeostasis of the proteins of the plasma for maintaining the volume and character of both the intravascular and extravascular fluid matrix of the organism, and for protecting the organism against loss of the essential part of the matrix, the blood, they have revealed also how much still needs to be learned. Here, as with other useful material, constancy is attained by storage, which stands between plenty and need, and in this respect the liver plays an important rôle. The sympathico-adrenal system seems to influence release from storage. Varying activity of the thyroid gland may also be determinative. Are special agencies required to manage the laying by of the reserves? We do not know.

References

Afanassiev. Pflüger's Arch., 1883, xxx, 385.
Boothby, Sandiford and Slosse. Ergebn. d. Physiol., 1925, xxiv, 733.
Foster and Whipple. Am. Journ. Physiol., 1922, lviii, 393, 407.
Luck. Journ. Biol. Chem., 1936, cxv, 491.
Meek. Ibid., 1912, xxx, 161.
Riecker and Winters. Proc. Soc. Exper. Biol. and Med., 1931, xxviii, 671.
Seitz. Pflüger's Arch., 1906, cxi, 309.
Stübel. Ibid., 1920, clxxxv, 74.
Tichmeneff. Biochem. Zeitschr., 1914, lix, 326.
Thomas. Arch. f. Physiol., 1910, 249.
Whipple, Smith and Belt. Am. Journ. Physiol., 1920, lii, 72.

VIII

THE HOMEOSTASIS OF BLOOD FAT

I

FATS, emulsified in minute droplets, and also fat-like substances, cholesterol and lecethin, are found regularly in blood in a concentration which, according to Bloor, differs greatly with different species, but is fairly constant for the same species of animal. Like carbohydrates, fats are composed of carbon, hydrogen and oxygen, but they have much less oxygen (i.e., relatively much more carbon and hydrogen) than carbohydrates. The phospholipins, e.g., lecethin and cephalin, contain not only these three elements, but also nitrogen and phosphorus. Lecethin is the best known of the phospholipins.

After a meal rich in fat there is an increase in the fat content of the blood; indeed, if the blood is drawn and allowed to stand without clotting, the fat droplets will rise and make a top layer like the cream on milk. Later after such a meal the lecethin content increases. About 70 per cent of this phospholipin consists of fatty acids. It is a non-toxic compound which can mix with water and which, consequently, is capable of following freely the courses of water transport. Bloor has suggested, therefore, that lecethin is the form in which fat is transferred from one region to another in the body. Evidence for this view has been presented by Sinclair who found, after administering a readily recognizable fatty acid, that it soon constituted 25 per cent or more of the fatty acids in lecethin. The manner in which this phospholipin is augmented in the blood during a period of increased transport of fat is not known.

II

Fat has various uses in the body, e.g., for the production of milk and as a source of energy. The relatively larger proportion of carbon and hydrogen in its composition makes its energy content more than twice that of an equal weight of carbohydrate. It is a concentrated source of heat, and can be used also by the organism for the performance of muscular work.

Fat, as everyone knows, can be stored in the body as adipose tissue. We have already noted that carbohydrate can be changed to fat and that this fat can be set aside in adipose tissue as a more compact reserve than glycogen. Adipose tissue is a modified connective tissue. It is found under the skin, at the back of the abdominal cavity around the kidneys, in the omentum (an apron-like fold of a membrane attached to the stomach) and also between the muscle fibers. A drop of fat almost fills each of the innumerable cells of the adipose tissue, so that the original structure of the cell, crowded to the outside, becomes a thin-walled bag stuffed with fatty matter. Some fat is found also in muscle cells, and in the cells of the liver, especially after much fatty food has been eaten. It is clear, therefore, that fat is stored by segregation, and not by inundation.

What leads to the storage of fat in large amounts—sometimes in grotesquely large amounts—in some persons, and in slight amounts in others, is not well understood. We know that, when there is deficient function of the thyroid gland, generally distributed and extreme obesity may result. Face, neck and shoulders, body, arms and legs, all may be swollen with the collections of subcutaneous fat. We know also that a slight scratch in a particular small region on the under surface of the brain can be used experimentally to produce adiposity, an effect which a tumor or other injury of this region will produce in man. Grafe has cited instances of *one-sided* overgrowth or undergrowth of fatty tissue in human beings, and has suggested that the regulation of fat deposit and release is under control of the sympathetic nervous system, and that the center for the control is in that part of the brain where damage produces adiposity. By a method which I shall describe later Newton, Moore and I have removed the sympathetic system of one side in kittens and have allowed them to live until they doubled their weight, but we have not seen any difference in the amount or distribution of the fat on the two sides of the body.

A large deposit of fat in the livers of rats was produced by Best and Channon on feeding a fat-rich, protein-poor diet, lacking choline. In two weeks the livers contained about 24 per cent of total fat instead of the normal 3 to 4 per cent. The presence of two or three milligrams of choline per day in the diet reduced the fat content to 10 per cent. Fat is stored abundantly in the livers of depancreatized dogs, as Chaikoff and Kaplan have shown, by prolonged treatment with insulin. A similar accumulation in the livers of rats has been found by Best and Campbell on administering extracts of the pituitary gland. Insulin does not act through the pituitary, however, for it produces fatty livers in depancreatized dogs after pituitary removal.

By feeding to rats a carbohydrate diet supplemented by heavy hydrogen (deuterium), Schoeheimer and Rittenberg found that fat depots accumulated fatty acids containing deuterium until a maximum was reached in six to eight days. The increase was due to synthesis of new acids. When the rats were transferred to a carbohydrate diet alone the fatty acids containing deuterium disappeared as rapidly as they had been synthesized. These experiments seem to indicate that stored fat may be much more mobile than has been commonly supposed.

III

If the regulation of fat storage is obscure, the regulation of its release is even more so. When fat is needed—in starvation, for example—for maintaining the energies of the body, it is removed from adipose tissue until the fat cells are practically empty. The constancy of the percentage of fat in the blood during many days of relative or complete starvation indicates that there is some governing agency which brings the fat from storage into the blood stream. How that is done we do not know.

There is some evidence of reversible action of fat-splitting enzymes, agents which would work automatically to store fat when the level in the blood rises and to release it when the level falls. It has been suggested also that the change from stored neutral fat to mobile lecethin may be favored by choline, as in the experiments mentioned above. Choline is the nitrogenous base of the lecethin molecule and therefore could participate in such a change; just how is not explained.

Another possible agent for mobilizing fat was disclosed by Himwich and his collaborators in 1931. They found that adrenalin or secreted adrenin greatly increased within a short time the plasma fatty acids. These results have been criticized by Long and Venning because the method used was defective. Employing a reliable method, Jones and Fish, in a study on man, observed about a half-hour after an intramuscular injection of adrenalin a typical moderate rise in the plasma fatty acids. The adrenal medulla may, therefore, play a rôle in the homeostasis of blood fat. But much more information than we now possess is needed before we shall understand fat storage and transport.

References

Best and Campbell. Journ. Physiol., 1936, lxxxvi, 190.
Best and Channon. Biochem. Journ. 1935, xxix, 2651.
Bloor. Physiol. Rev., 1922, ii, 106.
Cannon, Newton, Bright, Menken and Moore. Am. Journ. Physiol., 1929, lxxxix, 84.
Chaikoff and Kaplan. Journ. Biol. Chem., 1935, cxii, 155.
Grafe. Oppenheimer's Handbuch der Biochemie, 2nd Ed., Jena, 1927, ix, 68.
Himwich and Fulton. Am. Journ. Physiol., 1931, xcvii, 533.
Himwich and Spiers. Ibid., 1931, xcvii, 648.
Jones and Fish. Journ. Clin. Investigation, 1935, xiv, 143.
Long and Venning. Journ. Biol. Chem., 1932, xcvi, 397.
Meigs. Journ. Biol. Chem., 1919, xxxvii, 1.
Schoenheimer and Rittenberg. Journ. Biol. Chem., 1936, cxiv, 381.
Sinclair. Journ. Biol. Chem., 1936., cxv, 211.

IX

THE HOMEOSTASIS OF BLOOD CALCIUM

I

WHEN we were considering the storage of sodium chloride by inundation I stated that we should take up another mineral constituent of the blood, calcium, at a later stage because it was managed quite differently. Calcium is, in fact, stored by a special sort of segregation. It deserves particular consideration, also, because it has so many and so diverse uses in the organism—for the growth of the skeleton and the teeth, for the repair of broken bone, for the maintenance of proper conditions of responsiveness in nervous and muscular tissues, for the coagulation of blood, and for the production of serviceable milk.

Calcium occurs in the blood in two forms, ionized or combined with protein. The normal content is about 10 milligrams per cent. Considerable variations from that concentration may be dangerous. If blood calcium is lowered by taking it out of solution, twitchings and convulsive movements follow. They can be quickly relieved by injecting enough of a soluble calcium salt to restore the proper percentage. Removal of the parathyroid glands, four small structures near the thyroid gland in the neck, reduces the blood calcium to less than 7 milligrams per cent, without any change in the content of sodium or potassium. As the concentration approaches 5 milligrams per cent convulsions appear. Calcium given intravenously or by mouth abolishes these tetanic contractions, but when it falls again they recur. It has been suggested that the disorder results from a reduced ratio of calcium to phosphorus in the blood, and recovery is due to restoration of the proper ratio when the calcium content is increased.

When blood calcium is abnormally increased the danger which develops is consequent on profound changes in the consistency of the blood itself. By repeatedly injecting an extract of parathyroid glands Collip was able to raise the calcium concentration from 10 up to about 20 milligrams per cent. With this doubling of the calcium content the blood phosphates were also doubled; and the non-protein nitrogen as well as the urea nitrogen were multiplied fourfold. The osmotic pressure was much augmented. And the blood became so viscous that it could hardly be made to circulate. These serious results of increasing the calcium and phosphorus content of the blood do not occur in the ordinary operations of the body; indeed, the precision of homeostasis of calcium in the blood is astonishing. If a calcium compound is injected intravenously it rapidly disappears—not excreted but deposited in the body. Bone salts are continuously on the move; even in starvation some are excreted, but the level in the blood remains unchanged. This constancy, obviously, is a protection against the dangers mentioned above, and it is, therefore, a provision of primary importance.

II

Although calcium is obviously needed by all of us in continuous excess during the period of life when the bony framework of the body is being formed, there are times in the experience of a woman when the demand for calcium is especially great. During pregnancy she must provide calcium for the developing fetus, and throughout the months of nursing she must provide an even greater amount in the milk, in order to give the baby the calcium requisite for growth. If the daily calcium intake of the mother is not sufficient to meet these requirements, it is removed from the osseous structures of her body. In these circumstances there may be specific "softening" or decalcification of the teeth. It is of great interest that in these processes the calcium of the blood is kept at a fairly uniform level.

As in the homeostasis of other materials that of calcium is made possible by storage—in this instance, storage by segregation—built up in times of plenty and taken out and used in times of need. Where are the reserves of calcium held? The special studies of Aub and his collaborators on cats and rabbits have demonstrated that the trabeculae in the bone marrow are easily made to disappear by a long continued diet deficient in calcium, and that they are readily restored by feeding a calcium-rich diet. Rabbits and cats were given a high diet of calcium or a low diet and at the end of some months the left fore-leg was amputated at the shoulder; the diets were then reversed. After a similar period the animals were killed, and in each instance the remaining upper long bone of the leg (the humerus) was

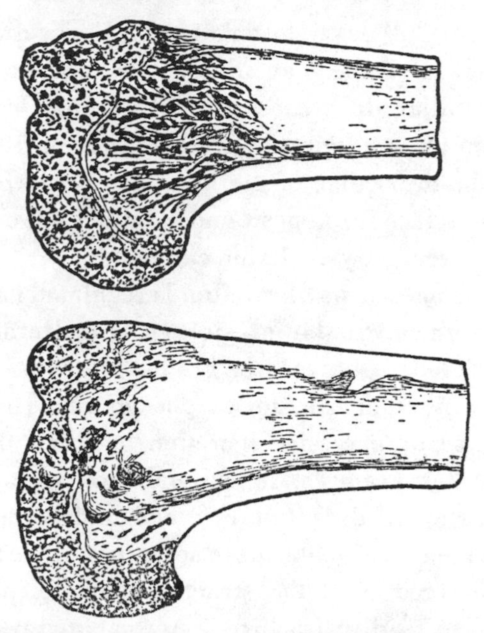

Fig. 25. Drawings from a photograph showing the effects of diet on the trabeculae in the humeri of a cat. The upper humerus, removed after a high calcium diet, contains many more trabeculae than its opposite member which was taken after a low calcium diet.

taken for comparison with its fellow. Whether the animal was placed first on a high diet or on a low diet, the results were always the same; the bone removed after high calcium feeding had many fine spicules or trabeculae projecting out into its cavity, and its opposite mate, after low calcium feeding, had few trabeculae. Figure 25, drawings from a photograph in a paper by Bauer, Aub and Albright, shows the characteristic differences which I have just described. The drawings represent the humeri of a cat; the animal was on a high diet for 80 days (upper drawing), and then on a low diet for 369 days (lower drawing). The difference in the numbers of trabeculae in the two conditions is very striking. Similar differences are observable in the other cases. From these admirable investigations it becomes evident that the trabeculae of the long bones, which present a very large surface for deposit and solution, serve as a storehouse of conveniently available calcium.

How the homeostasis of calcium is regulated has not been fully determined. So far as storage is concerned the reserves in the long bones may be gradually depleted by a diet inadequate in either calcium or phosphorus. They may be depleted also by excessive secretion of either the thyroid gland—as, for example, in exophthalmic goiter—or the anterior portion of the pituitary gland. Indeed, under these conditions the spaces in the more solid part of the bones may become enlarged until the structure appears porous and the bones are readily fractured. Accompanying this drain on the calcium reserves is an abnormally high loss from the body through the kidneys, although the concentration in the blood is maintained within limits of normal variation.

Evidence indicates that the parathyroid glands are the most important agents in the organism for preserving calcium homeostasis. Deficient secretion of these glands, like their removal, is associated with reduced blood calcium; the phosphorus concentration, however, is raised. As a consequence of this shift of the two levels tetanic muscular contractions appear, just as in the removal experiments. If parathyroid secretion is excessive, blood calcium is increased, blood phosphorus is lessened, and the discharge of both substances in the urine is augmented. The extra calcium in blood and urine is taken from the bones, and unless food rich in calcium and phosphorus is eaten, the bones may become softened by decalcification and the site of multiple cysts.

Furthermore, if the four parathyroid glands are lacking or inadequate there is defective deposit of dentine in rodents with continuously growing incisor teeth, and defective laying down of supporting bone around a fracture. After parathyroid tissue is implanted in a rat from which the glands have been removed, Erdheim found that the process of deposit of normal dentine in the incisors is restored. Again, in pregnancy and lactation, when an active passage of calcium from mother to offspring is proceeding, there is a special parathyroid overgrowth. Evidently these four small structures are intimately concerned with the government of calcium metabolism. Ordinarily storage and release of that important element are controlled with admirable exactness, so that a remarkably constant homeostatic condition is preserved. The mode of functioning of the parathyroids—whether they respond directly to slight variations of homeostasis or are stimulated by nerves or hormonal agents—is not clear. Lamelas found in the cat that blood calcium remains fairly uniform after either exclusion or increase of adrenal secretion. Possibly other glands of internal secretion, notably the anterior lobe of the pituitary body, are influential; Friedgood and McLean showed that repeated injection of an extract of this lobe significantly increased the calcium concentration in the blood—a result which they interpreted as due to hormonal stimulation of the parathyroids.

Possibly it is not out of place to note that vitamins in the food may influence the calcium stream through the organism. Vitamin D, prominent in cod-liver oil, facilitates absorption from the intestine and storage in the bones. In very large doses it has the same effects as large doses of parathyroid extract. And when it is deficient in the diet a softening of the bones, as in rickets, may appear in either children or adults. The suggestion has been offered that vitamin D has its influence through action on the parathyroid glands, but the supporting evidence is not conclusive.

We have now reviewed the evidences of constancy of glucose, sodium chloride, proteins and calcium in the blood. These include the chief agencies which determine its osmotic pressure. Measurement of the osmotic pressure of the blood has been made by Margaria on 18 men and 16 women, between 20 and 45 years of age. The men represented nine different nationalities. All subjects rested for a few minutes before the blood was taken (from an arm vein), but there were no precautions regarding meals, time of day, previous exercise, etc. The osmotic pressure was found to be amazingly uniform; the probable individual deviations were for men only 0.5 per cent and for women 0.6 per cent.

References

Aub. Journ. Am. Med. Assn., 1937, cix, 1276.
Aub, Bauer, Heath and Ropes. Journ. Clin. Invest., 1929, vii, 97.
Bauer, Aub and Albright. Journ. Exper. Med., 1929, xlix, 145.
Collip. Journ. Biol. Chem., 1926, lxiii, 395.
Erdheim. Zeitschr. f. Pathol., 1911, vii, 175, 238, 259.
Friedgood and McLean. Am. Journ. Physiol., 1937, cxviii, 588.
Lamelas. Am. Journ. Physiol., 1930, xciii, 111.
Margaria. Journ. Physiol., 1930, lxx, 417.

X

THE MAINTENANCE OF AN ADEQUATE OXYGEN SUPPLY

I

THE cells of our bodies are more closely dependent on oxygen than on any other substance which we obtain from the outer world. Professional fasters have proved that we can live without food for weeks, using up the bodily reserves of glycogen, fat and protein, and finally resorting to the actual structure of our own muscles and glands, without suffering any detectable permanent injury. Men who have been exposed to privation have proved that we can also go for days with no water intake; again, no doubt, drawing upon the stored water of the organism. But the conditions are different in oxygen want. There are important nerve cells in the brain which, if totally deprived of oxygen for more than eight minutes, undergo such profound destructive changes that they do not recover. This peculiar difference can be reasonably related to the fact that in the human economy no storage of oxygen has been provided. Food and water may not be always available, and stocks are accumulated in times of plenty for use in times of shortage. Oxygen, on the other hand, is always present as about 21 per cent of the great ocean of air all about us. We can take it whenever we need it. Storage is therefore unnecessary.

Circumstances arise, of course, when the oxygen delivery may be impaired. We have learned something of one such circumstance in hemorrhage. Ascension by airplanes or in mountain climbing to high altitudes, where the pressure of oxygen in the atmosphere is much reduced, offers another instance. Poisoning by illuminating gas or automobile exhaust—a state in which the place of oxygen in the circulating red blood corpuscles is usurped and tightly held by carbon monoxide—also produces a relative anoxemia. But these may be regarded as more or less unnatural conditions.

Continuous activities of the organism, such as the beating of the heart, the rhythmic movements of respiration, the persistent slight contractions of our skeletal muscles, require a continuous delivery of oxygen to permit the burning of the waste products of those activities. Whenever the activities are increased, as in vigorous muscular exertion, the minimal supply of oxygen, quite adequate for the resting state, would fall so far below the new need that the exertion could not continue. The machinery would quickly be compelled to stop because clogged by the accumulation of waste. Homeostasis is achieved in this situation by speeding up the continuous process.

II

The oxygen requirement of a man of average size, while he is at rest, may be only 0.25 to 0.30 liter (quart) per minute. When he changes to very severe muscular effort it may rise to 15 liters per minute or higher. Even in the most favorable circumstances, however, the maximal amount of oxygen which can be taken into the body and used is at the rate of 4 liters per minute. Thus during highly strenuous exercise the intake of oxygen may be from 10 to 12 times what it is during rest and yet be far short of what is actually needed at the time.

When this situation arises the lactic acid which develops as an accompaniment of muscular contraction cannot be burned to carbon dioxide or turned back into its precursor, glycogen. Consequently it accumulates in the muscles. Contraction can continue, but with decreasing efficiency, because the increasing concentration of the acid interferes with the contractile process. Evidently, great exertion does not depend on immediate oxidation. The lactic acid which we produce during a violent effort—for example, during a 225 yards run—is temporarily neutralized in the muscles and in the fluid matrix, and only later is burned or transformed into its precursor. Thus we may do work far beyond the energy supply represented by the oxygen delivery at the time. But we accumulate lactic acid. That must be got rid of. Some of it, which is present in the blood as sodium lactate, is passed out through the kidneys. But almost all of it is burned or reconverted into glycogen. This burning occurs after the exercise has ceased. We therefore borrow the ability to go on working beyond the limit set by the oxygen available, but only on condition that we take in enough oxygen later to burn the accumulated waste. We thus run up what Hill has called an "oxygen debt." We pay the debt during the continued deep respiration which, after we have stopped making any exertion, persists perhaps half an hour or more. In a man who ran 225 yards in 23.4 seconds, normal quiet breathing did not return until 27 minutes after the end of the run. The extra oxygen used during that period, above that used in the quiet state during a similar period, measured the amount of the debt.

From the foregoing considerations it is clear that in the delivery of oxygen to the remote cells of the organism homeostasis is provided only when the activity is moderate. Whenever supreme exertion or prolonged vigorous exercise is engaged in, the adjustments of the respiratory and circulatory systems are not fully adequate to prevent development of excess acid in the fluid matrix, and then devices to maintain a neutral reaction must be brought into play. The present chapter on the supply of oxygen and the next chapter on the neutrality of the blood are therefore closely related.

III

To supply the need for extra oxygen during muscular work the functioning of the lungs, the heart and the blood vessels is modified in various and complex ways, each directed towards providing an amount of oxygen sufficient to meet the requirements of the laboring parts or to pay the oxygen debt if the requirements have not been met during the period of labor.

The respirations, first of all, become deeper and more frequent. This change occurs at the very start of a muscular effort, too soon to be caused by any other agency than the nerve impulses from the cerebral cortex which initiate the effort itself—a bye effect produced automatically as an incident of the main action. Thereafter the greater volume of breathing—the deeper and faster respiratory movements—that we have all noted when exercising vigorously, is due to an increase of carbon dioxide in the blood. This increase results primarily from the larger output of carbon dioxide (CO_2) from the working muscles, where lactic acid is being produced and is being burned to CO_2. The higher concentration of the CO_2 in the muscle cells than in the lymph gives it a higher diffusion pressure (i.e., a greater tendency to spread), and therefore leads to its diffusion into the lymph, and thence, for a similar reason, to its diffusion into the blood flowing through the capillaries. The blood carries it to the lungs.

In the lungs the blood is spread out in a net of fine capillaries lying in the extremely thin walls of little air sacs or alveoli. These are the tips of the smallest twigs of the respiratory tree, whose trunk is the wind pipe or trachea, and whose branches are the bronchial tubes. It has been estimated that if all the air sacs in the lungs of a man were flattened out in a continuous sheet, a surface of about 90 square meters, or about 1100 square feet, would be produced, and that more than 800 square feet of it would be occupied by the pulmonary capillaries. It is evident that a very large area is provided for the exchange of gases between the blood and the alveolar cavities.

The percentage of carbon dioxide in alveolar air is about 5.25. When an extra amount of carbon dioxide is brought to the lungs in the venous blood it has a higher diffusion pressure than that in the alveoli and it therefore passes out of the blood into the alveolar cavities. This process goes on, however, only until a balance has been established between the tendency of the gas to enter the lungs from the blood and the tendency in the opposite direction. In this process the percentage of CO_2 in the alveoli is raised, of course, and consequently not so much of the gas can escape from the blood as in normal, quiet existence. The blood passes through and out from the lungs, therefore, carrying a larger load of CO_2 than usual. This arterial blood, with its increased diffusion pressure of carbon dicxide, now visits, among other places, the "respiratory center" in the medulla oblongata at the base of the brain. There, either because the gas diffuses from the blood into the cells of the center or because the higher pressure in the blood prevents the carbon dioxide, which results from the activity in these cells, from diffusing out, the gas content of the cells increases. This acts as a stimulus. The cells now discharge nerve impulses to the muscles of respiration so that their action is much more vigorous than before. The greater ventilation of the lungs naturally results in pumping out the extra carbon dioxide in the alveoli. When that has continued to the extent of reducing the percentage to the normal level of calm existence, the extra stimulation of the respiratory center subsides and quiet breathing is resumed. No better example of the automatic regulation of a continuous process, so that relative constancy of the fluid matrix is preserved, can be cited than this delicate agency in the brain, which, by keeping uniform its own status, protects all the rest of the body against the accumulation of acid waste. Some indication of its sensitiveness can be seen in the fact, demonstrated by Haldane and Priestley, that an increase of only 0.22 per cent of the carbon dioxide of the alveolar air will cause the ventilation of the lungs to be increased by 100 per cent!

In vigorous muscular effort the pulmonary ventilation in man may be increased from 6 liters per minute to 60 liters or more. That means that the passage of air in and out of the bronchial tubes must be very much more rapid than before, with the development of much more friction. In the walls of the bronchi are ring-like muscles which govern the size of the tubes. These muscles, like those in the arterioles, are controlled by nerves from the sympathetic system. By use of the denervated heart, in the manner already described, Britton and I showed that even in slight muscular movements the sympathico-adrenal system is brought into action. In figure 26 is shown the increase in rate of the denervated heart when an animal simply walks across the floor. In this case the rise was 20 beats. In 27 cases the average increase of the pulse rate after such activity was 17 beats per minute. This was due, in these circumstances, to adrenal secretion alone, for, as you note in figure 26, after the adrenal glands were rendered inactive the same muscular movement caused only an insignificant increase of two beats. Observe that when the animal struggles the sympathico-adrenal involvement is much greater, for the rate of the denervated heart rose much higher. In a large series of observations on different animals the average increase was 49 beats per minute. The effect of the sympathico-adrenal system on the bronchioles is to cause them to *relax,* an effect well illustrated by the dramatic relief which is experienced by the sufferer when adrenalin is administered in the treatment of bronchial asthma. It seems altogether probable, therefore, that during great

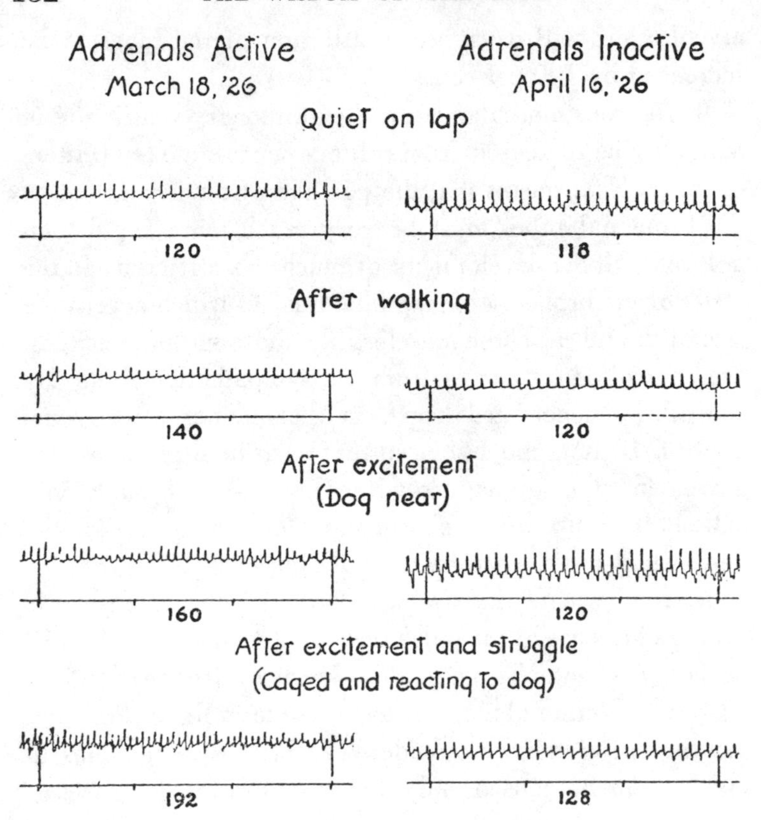

Fig. 26. *Original records of the rate of the denervated heart of cat 27 taken before (on March 18) and after (on April 16) secretion of adrenin was excluded (on April 6). Time in five-second intervals. The basal rate with the animal quiet on the lap is to be compared with the rate after walking, after being excited by the presence of a dog, and after being caged and reacting to a barking dog.*

muscular exertion the bronchioles are more dilated than during rest. The dilation would present larger passageways for the to-and-fro movement of the large volumes of air required in vigorous effort, and would thus lessen the frictional work that has to be done.

An additional beautiful feature of the economy of the respiratory process is that as the deeper breathing prevents the piling up of carbon dioxide in the lungs—and consequently in the blood, as well—by pumping it out, the deeper breathing pumps in, at the same time, extra oxygen. By the operation of this admirable automatism the percentage of oxygen in the alveolar air, and hence its diffusion pressure into the blood, is maintained at or slightly above the normal level, although there may be a five- or tenfold increase in the use of oxygen in the vigorously active body.

IV

Because of the double service of the respiratory movements the blood is enabled to unload its volatile waste, carbon dioxide, and to take on promptly as it passes through the pulmonary capillaries its load of oxygen. To understand the adjustments in the circulatory system which lie back of these transfers of the respiratory gases in the lungs, we must remember that the carriage of oxygen and carbon dioxide is dependent on the red corpuscles of the blood and that, although their number can be increased in emergencies, it is nevertheless limited. In such conditions, as we have already learned in our survey of the compensations which occur after severe hemorrhage, the only way to increase the carriage of these gases is to increase the use of the carriers; in other words, to multiply the number of trips which the carriers make between the lungs and the active parts. This in fact takes place, but in addition the processes of loading and unloading are facilitated at the two stations. We shall now consider these adjustments in detail.

First, in order that there shall be a larger output of blood from the heart per minute there must be a larger return of blood to the heart through the veins. This effect is achieved by a variety of alterations in the circulatory system when we engage in muscular exercise. The vasomotor nerves which govern the size of the blood vessels in the capacious vascular area of the stomach and intestines cause these vessels to contract. In consequence much of the blood is driven out of them and into the vessels of the muscles, which, as we shall see, have a greatly enlarged capacity when the muscles are at work.

Now the muscles, which in labor are contracted more or less rhythmically, exert a more or less rhythmic pressure on the vessels, especially the small veins, which lie within and between the muscle bundles. Since there are valves in the veins which permit a flow of blood only in the onward direction, towards the heart, the rhythmic pressure necessarily promotes that flow. This effect can be readily demonstrated; with the right hand, grasp firmly the left wrist, then rapidly and repeatedly clench and relax the left hand and note the quick filling of the veins on the back of that hand. Another type of pumping action on the veins occurs in the functioning of the great dome-shaped muscle of respiration, the midriff or diaphragm, which separates the thoracic from the abdominal cavity. When this muscle contracts it becomes less dome-shaped, i.e., somewhat flatter at the sides, as it presses downwards on the abdominal contents. Thereby it increases the pressure on the great vein, the lower vena cava, which leads blood upward through the abdomen from the legs. Here again valves in the branches of the vena cava prevent the backward flow of the blood into the legs. The increased abdominal pressure, therefore, favors the onward flow towards the heart. But conditions in the thorax are also favorable to this end. The lungs, as they lie in the chest cavity, are in a stretched state; they press inward on the air which they contain. The structures outside the lungs in the chest cavity, e.g., the veins, are therefore subjected to less than atmospheric pressure. With each inspiratory act the thorax is enlarged, the lungs are stretched still more and the pressure on the blood vessels is consequently still less. Thus at the same time that the abdominal pressure is increased the thoracic pressure is decreased, and not only is the flow from the leg region still further promoted, but the flow from the arms and head as well. During the next expiratory act the returning venous blood, of course, accumulates in the veins outside the chest—in the arms, the neck and the abdomen. But at the next inspiration the conditions just described recur, and the accumulated blood is driven to the heart. Thus, to the pumping action of the limb muscles is added the pumping action of the diaphragm as a factor promoting a greater utility of the blood.

Note the nice economy of this organization in the body. The contracting muscles which need extra oxygen because of their contractions, automatically favor the securing of the needed oxygen by returning the blood which carries it. And the diaphragm, which, as we have seen, is made to pump more vigorously during exercise, not only maintains in the lungs the oxygen supply for loading the oxygen carriers, but also aids to speed up the circulation of the carriers and thereby to augment their delivery of oxygen to the needy tissues.

V

The heart cannot put out more blood than it receives, and it must keep pace with the blood that is returned or there will be back pressure and stagnation in the veins. In short the output of the heart is governed by the volume of venous blood delivered to it. Since this volume is greatly augmented with the onset of muscular work the question arises, how will the heart deal with the task? Two methods are possible: the heart may expel more blood with each beat, and the rate of beating may increase. Both methods are employed.

The output of the heart per beat, or what Yandell Henderson calls the "stroke volume," must be determined by indirect measurements. The results obtained by different investigators, using various methods on relatively few subjects, have yielded figures which are consistent only in their general features. It appears to be clear that the heart beating at a slow rate, when the subject is at rest, does not nearly empty itself; the residue of blood in its ventricles after each contraction may be 25 per cent or more of the amount which was there before. When the heart relaxes in diastole the muscular walls are soft and flabby, especially on the right side where the venous blood is received. It yields, therefore, to the inrush of the blood delivered to it from the great veins, and becomes much more distended than usual. The English physiologist, Starling, discovered the highly interesting and important fact that within reasonable limits the more the muscle strands of the normal heart are extended the more powerfully they contract! Thus quite automatically the larger the amount of blood injected into the heart the greater the vigor with which it is ejected. The residue in the ventricles at the end of the contraction is therefore relatively less when the heart is beating energetically than when it is beating at a quiet pace. By reason of these adjustments the output per beat during hard work may be approximately twice what it is when we are at rest.

Merely doubling the output of blood per beat, and consequently per minute, obviously does not meet the demands of a situation in which the oxygen delivery must be increased ten fold or more. The limitations of the increase of the cardiac capacity is compensated for, as with the use of the red blood corpuscles, by a more rapid service. The rate of the heart beat, which ranges in the neighborhood of 70 per minute when we are quiet, runs up to 140 or higher when we exercise energetically. This addition to the doubling of the stroke volume of the heart greatly improves the possibility of satisfying the need for extra oxygen in muscular work.

The more frequent contraction of the heart is initiated and maintained by a variety of agencies. It is not necessary that work shall be actually in full swing before the heart begins to beat faster. In the adjustments of breathing we have already noted that the very act of making a motion is accompanied by an increase of respiration, because nerve impulses, attending this act, excite the respiratory center in the brain. Similarly when we start to move, the pulse becomes more rapid because vagal nervous influences, which are continuously holding the heart in check, are more or less suppressed. These are devices for prompt adaptation to need, that appear in two distinct systems of organs, which are, however, as must be evident, closely related in their coöperative functions.

The pump-like action of the limb muscles and of the diaphragm, that drives onward the venous blood into the right auricle and ventricle, causes an increased pressure in the veins, as evidenced by the prominence of the veins just beneath the skin during exertion. This increased pressure continues and accentuates the nervous effects just described, for when it is applied inside the heart it causes a greater distension of the right auricle and the adjacent endings of the great veins, and thereby starts a reflex, the so-called "Bainbridge reflex," which suppresses still more the vagus check on the heart rate, and hence the beat becomes and remains still faster.

The sympathetic nerves supplying the heart also are known to be excited by the increased venous pressure in the right auricle. They become active, furthermore, when muscular exertion is very strenuous, and especially if emotional excitement accompanies the effort, as in competitive games. We have already learned that the function of these nerves is to accelerate the heart beats. The actual records of the beats, reproduced in figure 26, demonstrate that even slight activity calls the sympathico-adrenal system into service. And the greater the activity, the greater the participation of that system. Furthermore, there is evidence that the nerve centers for sympathetic control may be influenced somewhat as the respiratory center is influenced; acidity may develop in them as a consequence of excess of carbon dioxide and the primary result of that may be stimulation. This suggestion is supported by the experiments of Mathison who showed that extra carbon dioxide in the respired air and therefore in the blood raises arterial pressure. Some observations made by the Linton brothers and myself likewise indicate that the waste products from muscular work bring into action the sympathico-adrenal system. We found, in confirmation of the Swedish physiologist, Johansson, that artificial stimulation of large muscle masses, quite isolated from nervous connections with the rest of the body, caused an increase of the rate of the denervated heart. This could be due only to chemical substances given off from the muscles into the circulating blood, because the only connection of the active muscles with the rest of the body was the blood stream. But in our experiments repetition of the same stimulus had no effect, or caused the heart rate to become somewhat slower, if the adrenal glands had been rendered incapable of action. The cardiac acceleration, evidently, was not a direct effect of the changes in the blood, but an effect produced by way of the sympathico-adrenal apparatus.

In vigorous muscular work, therefore, the remarkably close correlation between the adjustments of the respiratory and the circulatory systems to the needs of the organism may thus be explained; though both systems are started into faster action by impulses incidental to a voluntary movement, they are maintained in the performance of their extra task by the increased concentration of carbon dioxide in the arterial blood. Later they gradually return to their quiet routine functions, because their extra activity has resulted in reducing the carbon dioxide to the resting level.

All these influences working in harmony provide, first, for ample reception by the heart of the larger volume of blood driven back through the veins, and then for efficient delivery of the blood to the lungs where the deeper ventilation accommodates the greater exchange of the respiratory gases, oxygen and carbon dioxide, and finally for powerful discharge of the oxygen-laden blood into the vast "arterial tree."

VI

Accompanying physical exercise there is an increase of arterial blood pressure. We have already learned that the pressure in the arteries is a balance between the energy of the inflow into them and the resistance to the outflow from them. The largely augmented output from the heart, other conditions equal, would alone lift the blood pressure to a higher level. But there is evidence that also the peripheral resistance is increased, especially by constriction of the capacious blood vessels of the abdominal organs by action of the vasomotor nerves of the sympathetic system. These two factors, cardiac and vascular, working together, bring about a considerable elevation of the head of pressure which forces the flow through the capillaries. In tests made on a man riding a stationary bicycle the arterial blood pressure rose at the start from 130 millimeters of mercury to 180, and during the continuation of the exercise it remained high, between 165 and 170 millimeters, i.e., a pressure equivalent to a column of about 8 feet of blood instead of 5½ feet, the pressure when at rest.

It is worth noting that the higher tension of blood in the arteries develops before the exercise is actually going on. Just as respiration and heart rate are increased at the start of activity, as incidental indirect effects of the nerve impulses which cause the activity, so likewise a rapid and abrupt elevation of blood pressure at the start may be attributed to the same source. Indeed, according to Weber, the mere thought of taking exercise leads to the shrinking of the abdominal viscera and to the expansion of the limbs.

The value of the increased pressure we can best appreciate when we consider that there is a dilation of the arterioles and capillaries in the active muscles. If the arterial pressure were barely sufficient to keep the blood circulating with a speed adequate for the body at rest—just above the critical level, for example—a widening of the vessels in any particular region would offer such an easy way of escape for the blood into the veins that it would run through this wider channel and thus would leave other regions without a proper supply. The heightened arterial pressure during vigorous effort not only provides against any such failure of the delivery of blood to inactive regions, but it also assures a rapid flow in large volume through the dilated vessels of the active regions, where the need for the oxygen-bearing blood is greatest.

The dilation of the arterioles and capillaries in active muscles is one of the most remarkable emergency adjustments for bringing supplies to the cells and for carrying away their waste. Careful studies have shown that when a muscle is resting, many of its capillaries are not in use or that they take turns in carrying blood; some open here for a time and then close down so that no blood runs through them, while other near-by capillaries open and serve the neighborhood. Only the capillaries which contain blood are visible under the microscope. The Danish physiologist, Krogh, compared an active muscle of one side of the body with the corresponding muscle, inactive, of the other side and discovered the astonishing fact that in the muscle at work the number of open capillaries may range from 40 to 100 times the number in the muscle at rest. What causes the capillaries to dilate is not yet understood. Local lack of oxygen or increase of carbon dioxide has been suggested as the occasion for the opening of vessels—bad conditions which would in this way bring their own remedy. Or the dilation may result from action of a subtle substance resulting from the wear and tear of the muscle as it pulls.

However the capillaries may be opened, the great importance of their being opened should not be overlooked. I have previously emphasized the point that it is in the capillary region that the exchanges between the blood and the fixed cells occur. Here all the adaptations of the circulatory system to physical work have their significance. The blood is bearing sugar and oxygen which the laboring muscles require, and it can carry away the carbon dioxide and water which result from the burning that attends contraction. The nearer the flowing blood can be brought to the muscle cells in their need for both these services, the more efficiently will the muscular work be performed. The remarkable unfolding of the unused capillaries in muscles as the muscles enter into action assures intimate relations between the cells and the blood stream.

We may now complete the circuit of the adaptive changes

in the circulatory system. It is clear that when the muscles are rhythmically contracting and massaging the vessels within and between them they are pressing onward a greater volume of blood than is present when the muscles are idle. In short, the laboring muscles act as if they were outlying hearts, receiving more blood when they work and pumping that blood back to the central heart and to the lungs for refreshment and a new service.

VII

We have now noted that the demand for extra oxygen is met by perhaps a six-fold increase of pulmonary ventilation per minute, by approximately doubling the stroke volume of the heart, by doubling the heart rate, and by raising the arterial pressure so that the blood flows in greater volume per minute through the expanded capillarics of the region where activity requires a greater oxygen supply. Two other quite remarkable arrangements remain to be mentioned. The first is another illustration of the use of an accelerated process to preserve homeostasis. This is seen in the facilitation of the gas exchanges in the capillaries of the lungs and of the contracting muscles.

The circulatory adjustments just summarized are directed towards increasing the number of trips of the red blood corpuscles to and fro between lungs and muscles in a given time. Although the blood flow through the capillaries is much slower than anywhere else in the circuit—a device which, as already noted, gives time for exchange between blood and tissues—when the speed of the circulation is increased it may be increased in the capillaries as well as elsewhere. That would mean that less time would be allowed for the carriers to unload carbon dioxide in the lungs and take on oxygen, and to perform the reverse processes where the muscles are at work. The beautiful fact was demonstrated by Oinuma that increase of carbon dioxide and also increase of temperature hasten the unloading of oxygen. When the muscles repeatedly pull, therefore, and produce not only extra carbon dioxide but also extra heat, these new conditions force a more rapid and complete tipping out of the oxygen from the carriers. Even in moderate exercise the speed of liberation of oxygen may be doubled because of the influence of CO_2, and the influence of the rise of temperature in the active muscles must be added to that. At the same time the cells are using oxygen at a quicker rate and there is less oxygen in the lymph immediately in contact with them. This results in a steeper diffusion gradient from blood to lymph and consequently in a quicker diffusion. It is noteworthy that these are relations which vary together, so that the greater the muscular work, the greater the effects on the speed of dissociation of oxygen from the corpuscles and its passage to the cells. Thus the benefits of the faster and larger volume flow through the muscle capillaries are utilized in an adjusted fashion.

The more rapid the separation of oxygen the higher will be its diffusion pressure in the plasma, and therefore, in turn, the more quickly will it pass into the lymph and thence into the active cells. As a result of this chain of events the cells receive oxygen in accordance with their varying requirements.

In the lung capillaries the blood is exposed to a relatively high diffusion pressure of oxygen and a relatively low diffusion pressure of carbon dioxide, as these two gases are related in the pulmonary air sacs. The carbon dioxide consequently escapes through the extremely thin films, which constitute the capillary and alveolar walls, into the alveolar space. But it does not leave unassisted. The entrance of oxygen into the red corpuscles from the alveoli helps to force out the carbon dioxide more rapidly than it would otherwise go. At each turning point of the circulation, therefore, the gas which hustles the other out seizes the vacated place in the carrier for itself and holds it until, in due course, it also is hustled out. No more fascinating interplay of processes than this is found in any part of the organism.

When the blood leaves the lungs it carries away, even during the rapid flow in violent muscular effort, about 95 per cent of its full capacity of oxygen, or about 18 cubic centimeters of oxygen in 100 cubic centimeters of blood. In the quiet, resting state the venous blood, when it returns to the heart, may bring back 14 or 15 cc. of oxygen in 100 cc. Only 3 or 4 cc. have been released in the outlying capillaries. Because of the efficacy of the factors whose action we have just reviewed, the reduction of the oxygen load in the active regions is much increased, so that even the *mixed* venous blood on its return may contain less oxygen than 5 cc. per cent. Thus the increase in the number of trips made by the corpuscles from lungs to needy tissues must be multiplied by the much greater utilization of the oxygen which the corpuscles carry.

VIII

One more striking provision for assuring an adequate supply of oxygen in case of need is seen in the sudden rise in the number of red corpuscles during muscular exercise. The phenomenon, which Barcroft especially has made clear, is more prominent in lower animals than in man. In the horse the number per cubic millimeter may increase as much as 20 per cent or more as a consequence of five minutes of hard work. This remarkable change is the only adaptation of the organism to oxygen want that resembles the resort to storage, which I have described in the homeostasis of other supplies than oxygen. When muscular exertion is severe and prolonged, glucose, as we have learned, is mobilized from the liver stores and distributed by the blood for use wherever it is required. In considering the reaction of the organism to hemorrhage the spleen was noted as a store house where red blood corpuscles are held in a concentrated collection. If the muscles of the spleen contract, as they do in circumstances which evoke action of the sympathico-adrenal system, e.g., whenever there is oxygen-want, they squeeze out the contents. In the cat, exercise may result in a diminution of the weight of the spleen from 26 to 13 grams—in a discharge, therefore, of 13 grams of fluid especially rich in red corpuscles. These corpuscles, of course, at once become carriers of oxygen and carbon dioxide, at a time when their services are in demand.

The elaborateness of these multiple devices for preserving homeostasis of the oxygen delivery to the fixed and secluded cells of the organism measure the importance of keeping that delivery constantly adjusted to the need. In all moderate and prolonged labor the close coöperation of the numerous factors which participate in this necessitous service results in quite efficient maintenance of an adequate supply of oxygen. When labor is violent and extreme, however, the supply may fail to meet the demand. Then non-volatile lactic acid begins to accumulate in the cells and diffuse outward into the lymph and blood. The blood must not be allowed to alter its state of near neutrality between an acid and an alkaline reaction. The provisions for protecting the organism against that dangerous possibility we shall next consider.

References

Bainbridge. Journ. Physiol., 1914, xlviii, 332.
Barcroft. Ergebn. d. Physiol., 1926, xxv, 818.
Cannon and Britton. Am. Journ. Physiol., 1927, lxxix, 433.
Cannon, Linton and Linton. Ibid., 1924, lxxi, 153.
Henderson. Physiol. Rev., 1923, iii, 165.
Hill, Long and Lupton. Proc. Roy. Soc., London, 1924, xcvi, 442.
Johansson. Skand. Arch. f. Physiol., 1895, v, 59.
Krogh. The Anatomy and Physiology of the Capillaries. New Haven, 1929, 64.
Mathison. Journ. Physiol., 1911, xlii, 283.
Oinuma. Ibid., 1911, xliii, 364.
Starling. The Law of the Heart. London, 1918.
Weber. Arch. f. Physiol., 1907, 300.

XI

THE HOMEOSTASIS OF NEUTRALITY IN THE BLOOD

I

IN THE previous chapter I repeatedly mentioned the formation of lactic acid and also of carbon dioxide (which in water solution forms carbonic acid) as a necessary accompaniment of muscular work. Phosphoric acid and sulphuric acid are likewise produced in the body from the oxidation of the phosphorus and sulphur contained in protein food. In certain conditions of disease acid substances of a still different character appear. On the other hand, basic radicles, such as sodium, potassium and calcium, are ingested in large amounts, especially in vegetable food, or acid may be lost temporarily from the body by secretion in the acid gastric juice, and thus may bring about conditions which tend to render the blood alkaline. It is of the greatest importance to the existence and proper action of the cells that the blood shall not vary to a noteworthy degree either in the acid or the alkaline direction.

The "reaction" of the blood is measured by the concentration of the hydrogen (H) ions in the plasma. The hydrogen ion is an atom of the element, hydrogen, bearing an electric charge. When hydrochloric acid (HCl) is added to water, the water becomes acid because the HCl breaks up, or dissociates, into its constituent ions, hydrogen and chlorine. The degree of acidity depends on the number of hydrogen ions present in the solution. Similarly the degree of alkalinity of a solution depends on the concentration of the combined elements, hydrogen (H) and oxygen (O), which, bearing an electric charge, make the so-called "hydroxyl ion" (OH). Pure distilled water (H_2O) dissociates slightly into H and OH ions, which are necessarily equal in number. It is regarded as neutral, not because there are neither acid nor alkaline agents present, but because they are both present in equal degree. In pure water at 22° C. there is one gram by weight of ionic hydrogen in 10 million liters of water, or the hydrogen-ion concentration is $1/10$ million or $1/10^7$ or 10^{-7}. In pure water the hydroxyl-ion concentration is also 10^{-7}. Indeed, the product of the negative exponents in any aqueous solution is found to be always –14. It is customary now to avoid these negative index figures, and to express neutrality by the notation, pH=7. If the hydrogen-ion concentration is 10^{-6}, the hydroxyl-ion concentration will be 10^{-8}, and the solution is acid. In the reverse relation, naturally, the solution would be alkaline. Thus if the index figure is less than pH 7, H-ions predominate and there is an acid reaction; and if the figure is more than pH 7, OH-ions are in excess and the reaction is alkaline.

II

Blood has a slightly larger concentration of OH-ions than of H-ions, the index figure being approximately pH 7.4.

Even minor variations from this reaction, which is just on the alkaline side of neutrality, are dangerous. If the hydrogen-ion concentration rises, so that the figure changes only to pH 6.95 (i.e., barely over the line in the acid range) coma and death result. And if the hydrogen-ion concentration falls, an increase in alkalinity as trifling as that indicated by a shift from pH 7.4 to pH 7.7 in the index figure, brings on tetanic convulsions. The heart rate of the dog has been seen to decrease from 75 beats per minute to 50 when the shift was pH 7.4 to pH 7.0, and to increase from 30 per minute to about 85 when the shift was in the alkaline direction (pH 7.0 to pH 7.8). If these limits are exceeded the effects are disastrous; with an acid fluid the heart relaxes and ceases to beat, and with a more alkaline fluid it also ceases to beat but usually stops in the contracted phase.

These are only a few examples out of many which could be cited to show the very grave dangers of altering the chemical reaction of the blood. Within a narrow range of variation the nervous system will operate perfectly and the heart will go on beating continuously. In health, the variations from the normal reaction do not extend far enough beyond the close confines to impair the activities of the organism or to threaten its existence. Before such extremes are reached agencies are automatically called into service which act to bring back towards the normal position the disturbed state.

The complete account of the mechanisms by which the reaction of the blood is kept close to neutrality, in spite of conditions which operate to push the reaction away in one direction or the other, would require elaborate and detailed consideration of involved physico-chemical processes.

We shall regard only the simpler aspects of the mechanisms.

III

Dissolved in the blood plasma is a compound of the chemical elements, sodium (Na), hydrogen (H), carbon (C), and oxygen (O) which is present in three parts. This compound ($NaHCO_3$) is ordinary cooking soda, and is known chemically as sodium bicarbonate. Likewise in the blood plasma is carbonic acid. The symbol of carbonic acid, which results from dissolving carbon dioxide (CO_2) in water (H_2O), is H_2CO_3. The reaction of the blood is determined by the relation of H_2CO_3 to $NaHCO_3$ as they exist in the plasma—the carbonic acid providing the hydrogen-ions, and the sodium bicarbonate, by action of the sodium, providing the hydroxyl-ions. If the carbonic acid is increased, as in exercise, the plasma becomes more acid. It can be rendered more alkaline by reducing the carbonic acid. If, for example, we engage for a minute or two in excessive, voluntary, deep and rapid breathing, we can lower the concentration of carbon dioxide in the alveoli of the lungs and hence reduce its diffusion pressure. The diffusion pressure of the carbon dioxide in the blood now meets less opposition on arriving at the lungs and hence more CO_2 leaves, a process favored by the higher concentration of oxygen in the alveoli that has resulted from the excessive breathing. In experiments by physiologists on themselves this way of rendering the blood more alkaline has been carried nearly to the extent of producing convulsions.

If a non-volatile acid, such as hydrochloric acid (HCl)

or lactic acid, which we may symbolize as HL, is added to the blood, it unites with some of the sodium of the sodium bicarbonate and drives off carbon dioxide, according to the following equations:

$$HCl + NaHCO_3 = NaCl + H_2O + CO_2 \text{ or}$$
$$HL + NaHCO_3 = NaL + H_2O + CO_2$$

The NaCl is common table salt, a neutral, harmless substance. The H_2O and CO_2 form the familiar carbonic acid, which is volatile. The addition of the strong acid, HCl or HL, has, to be sure, made the blood temporarily more acid by increasing the carbonic acid. As we have already learned, however, the increase of CO_2 stimulates the respiratory center, and the consequent increased ventilation of the lungs quickly and readily gets rid of the extra acid—both that produced by displacement from $NaHCO_3$ as in the equations above and that which is now in excess because the $NaHCO_3$ has been reduced. As soon as the extra carbon dioxide is pumped out, and the usual ratio H_2CO_3 to $NaHCO_3$ gradually returns, the normal reaction of the blood is restored and the deeper breathing stops.

In violent muscular exertion we have seen that lactic acid may be produced so abundantly that not enough oxygen can be brought, at the time, to burn it to carbonic acid. If the lactic acid escapes from the muscle cells into the blood, it is first dealt with in the manner just described, i.e., by the formation of neutral sodium lactate. The lactic acid left in the muscle cells is likewise neutralized by the alkali set free there. But this is a limiting condition, which can be remedied only by burning the lactate to carbonic acid and water. There is an "oxygen debt" which must be paid. The extra oxygen taken in after the excessive exertion has stopped is used first to burn the lactate accumulated in the muscles. As this is burned, its concentration lessens; then the lactate in the plasma slowly diffuses back into the muscles and there in turn is burned. The CO_2 formed from the burning is the cause of the continued deep respiration while the oxygen debt is being paid. In part it is breathed away. In part, however, it remains in the blood. There it unites with the sodium which, because of the burning, has been set free from the compound, sodium lactate. And thus the sodium bicarbonate of the blood, that was decreased during the period of intense muscular work, is brought back to its normal concentration.

IV

In the circumstances described in the foregoing paragraphs the sodium bicarbonate of the plasma served to protect the blood from any considerable change in the acid direction. Because of its capacity to perform that function it is called a "buffer" salt. Another buffer salt present in the blood, especially in the red blood corpuscles, is alkaline sodium phosphate (Na_2HPO_4). When acid is added to blood, not only is it "buffered" by sodium bicarbonate, but also by the alkaline sodium phosphate, as shown in the following equation:

$$Na_2HPO_4 + HCl = NaH_2PO_4 + NaCl$$

Again note that common salt ($NaCl$) is formed, and also acid (dihydrogen) sodium phosphate. It happens that both the "alkaline" and the "acid" sodium phosphates are almost neutral substances. The strong hydrochloric acid, in the change symbolized in the equation, has, therefore, not altered the reaction of the blood to an important degree by transforming the alkaline into the acid form of the phosphate. The acid phosphate has, however, a slightly acid reaction and it must not be permitted to accumulate in the fluid matrix. Unlike carbonic acid, it is non-volatile and therefore cannot be breathed away. Here the kidney plays its part in restricting the oscillations of the acid and the alkali of the blood.

In the plasma the ratio of NaH_2PO_4 to Na_2HPO_4 is as 1 to 4. In such a relation these two salts are filtered through the glomeruli. As mentioned earlier, the base (e.g., Na) in a salt is held more uniformly in the blood than the acid radicle with which it is combined. The fixity of the base appears to be the more important condition for the organism. Now in passing through the kidney tubules the disodium phosphate is changed to the acid form, thus:

$$Na_2HPO_4 + H_2O = Na + OH + NaH_2PO_4$$

The base, Na, is reabsorbed, to some degree as bicarbonate, and the acid sodium phosphate is excreted. By this change in the kidney tubules the ratio of the acid to the alkaline phosphate, which at the start was as 1 to 4, becomes, at the end, in the urine, as 9 to 1, and thereby much acid may be eliminated. Whenever there is a tendency towards an acid reaction in the blood the acid phosphate is excreted in greater amount and thereby the reaction is corrected in the alkaline direction.

If large amounts of non-volatile—and consequently non-respirable—acid appear in the blood, there is danger that the fixed bases of the blood salts, especially sodium, may be carried away through the kidneys and thus lost from the body. In this condition, it is interesting to note that ammonia (NH_3), which is alkaline, can be used to neutralize the acid in place of sodium. Ammonia is a waste product of organic processes, which is ordinarily transformed into the neutral substance, urea, and eliminated. Whenever loss of the fixed bases, e.g., sodium, calcium and potassium, is threatened, ammonium salts are formed and discharged into the blood, then filtered out through the glomeruli and the kidney tubules.

V

A modification of the foregoing processes occurs when the blood tends to become alkaline. Let us suppose that a sharp pain or exposure to high temperature has caused unusually deep and rapid breathing. The carbon dioxide percentage of the air in the lungs is thereby reduced and in consequence it is reduced also in the blood. The ratio of H_2CO_3 to $NaHCO_3$ is lowered, i.e., the reaction shifts further to the alkaline side. There is free escape of carbonic acid from the respiratory center, and consequently the stimulus is lacking which induces the discharge of nerve impulses to the diaphragm and other muscles used in breathing. For a time under these conditions respiration may cease altogether. But in the absence of breathing, carbon dioxide, which is continuously being produced by the beating heart and other persistently active organs, accumulates in the blood until the normal ratio of H_2CO_3 to $NaHCO_3$ returns, whereupon the rhythmic ventilation of the lungs begins again. And if the reaction of the blood is for some time swung in the alkaline direction, alkaline sodium phosphate and other basic salts, with perhaps no ammonium salts, are excreted until neutrality is assured.

We have seen how hazardous are the alterations in the chemical reaction of the blood. Here homeostasis of the fluid matrix is of the utmost urgency. The very delicate balance between a dangerous increase of the hydrogen-ion concentration, and an almost equally dangerous change in the opposite direction is maintained chiefly by the extraordinary sensitiveness of the respiratory center in the brain. Coöperating with this center are the highly efficacious buffer salts in the blood itself, which absorb the first shock when there is a charge of acid into the blood. The kidneys, more slowly adapting their functions to the situation, prevent accumulation of either non-volatile acid or alkaline salts. We may think of these various agencies as sentinels, continuously on the alert, ready at the first indications of a change to act in such ways as to prevent a harmful swing away from that normal steady state of the blood which is near the neutral position.

Reference

Henderson. Blood. New Haven, 1928.

XII

THE CONSTANCY OF BODY TEMPERATURE

I

ONE of the most striking and easily observed constants of the internal environment is that of the temperature of "warm-blooded" animals. Although in normal human beings there is a daily swing from a low point about 4 A. M., when the thermometer readings average 36.3° C. (97.3° Fahrenheit) to a high point about 4 P. M., when they average 37.3° C. (99.1° F.), it does not vary much beyond this range. The constancy is so reliable that the thermometer makers can stamp "98.6°" on the Fahrenheit scale with assurance that it will mark closely the mean temperature of the healthy person everywhere.

The uniformity of our body temperature is not always maintained. Alcohol and anesthetics may abolish the regulatory processes, and then, on exposure to cold, heat is rapidly lost. Thus during alcoholic intoxication a man has suffered a fall of temperature to 24° C. (75° F.) and later has recovered a normal state. On the other hand, in the course of infectious diseases the fever may rise to 40° C. (104° F.) or higher without causing disability. But these variations bring dangers and limitations. For example, if the temperature rises to about 42°–43° C. (107°–

109° F.), as it may in sunstroke, and remains there for some hours, it produces profound disturbances in nerve cells of the brain. Also, 24° C. (75° F.) is much lower than is compatible with activity. As Britton has noted, the heart beats very slowly, respirations are shallow and infrequent, and deep lethargy prevails. There is a good reason, therefore, in avoiding the extreme variations.

There is good reason also in preserving the normal constancy of body temperature. Its value becomes clear as soon as we compare the influence of cold on ourselves and on lower animals, e.g., amphibia and reptiles, which have no heat regulatory apparatus. I have already mentioned the effect of low temperature on the frog. As the weather turns cold he becomes cold, too, and his actions are more and more sluggish. His heart beats rarely, and as he lies inert, deep in a frigid pool, he does not breathe at all. Thus he remains until he is warmed again. This behavior of the frog is chiefly due to the fact that many of the essential processes going on in organisms are chemical and that the speed of chemical processes varies with the temperature, an increase of 10° C. practically doubling the rate. The "cold-blooded" animals, therefore, having the temperature of their surroundings, can act with alacrity only when the weather is warm; the warm-blooded, which maintain a fairly fixed high temperature in spite of external cold, can act quickly at all times. By the preservation of constancy of the internal environment they are freed from the influence of vicissitudes in the external environment.

II

To understand the regulation of temperature in our bodies we must realize, first, that heat is being continuously produced by every variety of activity in which our organs engage. All the energy of the powerfully contracting heart is ultimately turned to heat inside us, for the mechanical work which the heart performs in building up the head of arterial pressure is spent in overcoming frictional resistance in the blood vessels. About three-fourths of the energy of our muscular activity appears necessarily as heat. And the processes in the liver and other glands are all accompanied by heat production. We have learned that when an organ becomes specially active, the blood flows through it in larger volume. The heat developed by the activity of the organ diffuses from the warm cells to the cooler blood. Thus the cells are prevented from becoming overheated, and the heat produced in one part is made serviceable in other parts. The man who swings his arms and stamps his feet vigorously on a cold morning is making heat in his muscles that the circulating blood renders generally useful for the cold regions of his body. An important function which the moving part of the fluid matrix performs, therefore, is that of equalizing the temperature throughout the organism. As we shall see later, it plays also an essential rôle in the management of heat loss through the skin.

There is evidence that heat production in the body is under control, and that in normal persons, under standard conditions, it is remarkably uniform. The standard conditions are satisfied by a fast of about eighteen hours after taking a mixed diet (commonly over night) and then a rest in the reclining position for about twenty minutes before the test. With the subject awake and lying quietly on his back, in a room temperature of about 20° C. (68° F.), the intake of oxygen and the output of carbon dioxide are measured during a series of short periods. From the figures for O_2 consumed and CO_2 discharged, the heat produced by the burning can be readily calculated. It is usually expressed in calories per square meter of body surface per hour or per day, and measures the so-called "basal metabolism." As we shall see later, there is a gradual diminution of the rate of heat production by the body as one progresses towards the years of senescence. Within limited periods of the existence of the organism, however, constancy of the basal metabolism is the rule. A man studied for six years by Benedict and Carpenter, at the Carnegie Laboratory in Boston, had a variation as little as 3.7 per cent each year from the average of all the years. In a dog tested by Lusk during two years the basal metabolic rate, in 17 observations, differed only 2.9 per cent. Such uniformity is astonishing.

We know that the metabolic rate is affected by disturbances of several glands of internal secretion; for example, the pituitary at the base of the brain and the cortex of the adrenal gland. But the gland that is most markedly and most directly influential is the thyroid, in the neck. When the thyroid is over-active, as in exophthalmic goiter, the metabolic rate commonly rises 50 or 75 per cent above normal, and cases of hyperthyroidism are known in which the rate doubled, i.e., the processes of heat production under standard conditions were actually going on twice as fast as in the healthy person. When the surgeon, in treating this state, removes a large part of the gland the proper metabolic rate can be restored. On the other hand, when the thyroid gland is defective or deficient, as in cretinism and myxedema, the rate of combustion may be from 30 to 40 per cent below the normal level. The metabolism of such patients can easily be raised to the normal level by feeding them thyroid gland or giving them an extract of it in substitution for the thyroid secretion which they lack.

What keeps the thyroid gland constantly active is not known. In cats which were studied by a group in the Harvard Physiological Laboratory, after different stages in the removal of the sympathetic nervous system, there was a slight drop in the metabolic rate when the nerve strands to the neck region were extirpated, but the effect was so slight as to be, perhaps, insignificant. The glands could not have been kept active in these circumstances by secretion of adrenin from the adrenal glands, because in certain cases these too were denervated and there was no clear difference on that account. For the present the control of the internal secretion of the thyroid must be left for further research. All that we know definitely is that the basal metabolic rate is one of the constants of the organism, that its constancy seems to be directly dependent on the proper functioning of the thyroid gland, and that the reliable uniformity of the basal metabolism is a condition for other phases of homeostasis, especially that of body temperature.

III

The homeostasis of body temperature, like that of the oxygen supply to the tissues, is achieved by modifying the speed of a continuous process. As we have seen, heat is continuously being produced by organic activity. Constant temperature can be maintained by increasing or decreasing the rate of heat loss or by increasing or decreasing the rate of its production, according to need. We shall consider first the agencies concerned in heat loss.

Let us suppose that conditions favor a rise of body temperature because, for example, a large amount of heat has been produced by very vigorous muscular work. In these circumstances the vasomotor nerves governing the size of the surface arterioles relax their grip, the vessels dilate, and the blood, warmed by the active muscles, flows in much larger volume through these arterioles and through the capillaries to which they contribute. In consequence the skin becomes red. If the surrounding air is cool the extra heat brought to the skin will pass out to it by radiation and conduction, and a rise of body temperature will be prevented.

If the outer air is too warm to permit the heat to pass to it, however, another process is invoked. Heat is lost by evaporation. When water evaporates, as much heat is taken from neighboring objects as would be required to cause the water to evaporate. This is a fact well known to persons who live in hot dry climates and who use porous earthen jars or canvas bags to cool the drinking water. The greater delivery of warm blood to the skin can be combined with the pouring out of sweat on the skin surface. Just as the evaporation of moisture from the outside of the porous containers cools the water within, so likewise the evaporation of sweat cools the skin and consequently the blood flowing through its capillaries. If the air is dry, large amounts of heat may be lost in this way. It is by this means that high external temperatures are withstood by stokers and by foundry workers who are exposed to the intense heat of open furnaces. Occasionally, however, persons are found with defective sweat glands. A man with that affliction, when exposed to the sun for a short time on a summer day, soon had a body temperature of 41.5° C. (nearly 107° F.). When such a person has to work hard in hot weather his only resource in avoiding the development of a fever is to wet his garments repeatedly and let the vaporization of water from them replace the vaporization of sweat. The very uncomfortable experience which we have on a day which is not only hot but also muggy is due to the high vapor density or humidity of the air, which interferes with the change of sweat to water vapor and thus prevents cooling.

To a considerable degree we lose heat also by evaporation of fluid from the surfaces of the respiratory tract. On a morning in winter we "see the breath" because the moisture added to the inspired air is promptly condensed when that air is breathed out to the cold surrounding atmosphere. On a hot day a similar evaporation is continuously going on and can be augmented by rapid respiration. As human beings we do not ordinarily use this process for cooling purposes alone, though in one case on record a man who could not sweat because of an atrophied skin and who breathed 6.32 liters of air per minute when he was at rest and his body temperature was normal, breathed nearly three times that much when his temperature rose to 39.9° C. (103.8° F.), and at the rate of 90 breaths per minute! In some lower animals—in the dog, for example—the quick movement of air to and fro in the respiratory passages, while panting, is the chief means of losing heat if the temperature tends to rise. In man, also, during and after vigorous exercise, the fast, deep breathing caused by excess production of carbon dioxide, has the nice quadruple effect of preventing the accumulation of carbon dioxide in the lungs, assuring the presence of plenty of oxygen there, pumping onward the venous blood and helping to get rid of the extra heat which results from the muscular activity.

As we have already noted, there is a considerable amount of heat produced by the organism as the inevitable by-product of existence. The basal metabolism, by definition, is the lowest degree of chemical oxidation when the body is at rest. Only by complete idleness can heat production be reduced to a minimum. When the surrounding temperature is high, therefore, not much advantage can be gained by restricting the development of heat. The main reliance must be the increase of heat loss by the methods which we have been considering.

IV

If the body temperature tends to fall, an interesting series of adjustments occur, all directed towards preservation of the steady state. First, heat which is being lost through the skin is conserved. To that end perspiration is reduced to a minimum. The surface vessels are contracted so that the warm blood from the interior is not exposed to the cold surroundings. And in animals provided with hair or feathers these appendages of the skin are lifted to enclose in their meshes a thicker layer of air, which is a poor conductor of heat. Of this last protective reaction only futile "gooseflesh" remains in us, and the single little hair standing upright in each hummock of the gooseflesh signals its futility. In place of the efficient protection which fur would afford, mankind has to resort to extra clothing—often the fur of lower animals!—to prevent too great loss of heat.

In addition to the constriction of surface vessels and erection of hairs, it is interesting to learn that a rise in the level of blood sugar occurs when the body is chilled. Erect hairs, constricted arterioles and hyperglycemia are signs that the sympathetic division of the autonomic nervous system is active. The question naturally arises, is secretion of adrenin, which is admittedly under sympathetic control, augmented when cold causes a discharge of sympathetic impulses? The answer to that question has important bearings, because adrenin not only collaborates with the sympathetic impulses which are diminishing the caliber of the surface vessels, but it has the power, as shown by Aub and McIver and their associates in the Harvard Physiological Laboratory, to accelerate the processes of combustion in all parts of the organism. Adrenin would have effects like those caused by opening the dampers of a furnace; burning would go on more rapidly. If, therefore, the adrenal medulla is made to secrete by chilling the body, the extra adrenin discharged would effect a faster production of heat just when there would be special need for it.

Microscopical studies of the adrenal glands by Cramer and others indicated that exposing animals to cold reduces the substance from which adrenin is derived. This was interpreted as evidence that cold stimulates adrenal secretion. But perhaps cooling retards or diminishes the production of that substance—the result would be what was observed. More direct evidence was required in order to determine the fact. Such evidence was furnished by Hartman and his collaborators in Buffalo. They made use of the completely denervated iris in cats. When these animals were wet with cold water, or cooled after being wet with warm water, the pupil dilated if the adrenal glands were functional, but not if they had been rendered incapable of action. The natural aversion of cats for water on the skin caused a certain amount of excitement, however, and the observed effect could not be sharply differentiated from the effect of excitement itself. To avoid this possible error and also for other reasons Querido and Britton, Miss Bright and I decided to carry on the investigation further.

In our experiments we again made use of the denervated heart, in unanesthetized animals, as an indicator of increased adrenin in the blood stream. We employed several methods of exposing the animals to cold. At first we held the animal, warm and comfortable, on a cushion near a window; and after the heart rate had been counted or registered we merely opened the window to the outer cold. The great advantage of this method was that the element of excitement and emotion was wholly eliminated because the surroundings were familiar; the only change was the opening of the window. In figure 27 is presented a copy of the original records of the effects on the rate of the denervated heart of exposing an animal thus to the cold air. Both adrenal glands were intact. The room temperature was 16° C. (60.8° F.), the outdoor temperature was –4° C. (24.8° F.). Observe that the basal rate was 118 beats per minute. Four minutes after the window was opened the

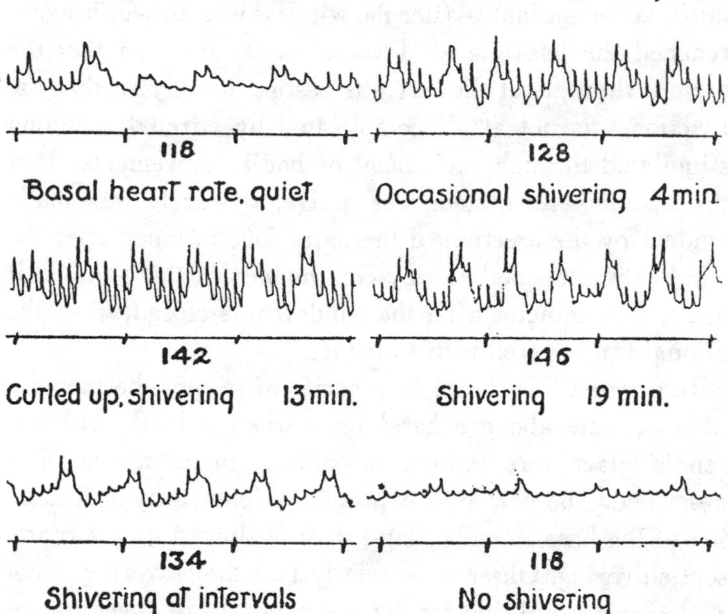

Fig. 27. Original records showing increases of rate of the denervated heart when cat 23 was exposed to air at –4°, January 29, 1926. Below the records taken while the window was open are written figures telling the time after the opening. Nine minutes after closing the window the rate had fallen from 146 to 134, and seven minutes later to 118 (the basal). Time in five-second intervals.

heart rate had increased 10 beats per minute. In thirteen minutes it had increased 24 beats. At that time the door was opened so that a strong draught of cold air rushed by. At the end of the next six minutes, that is, nineteen minutes after the start, the rate had reached 146 beats per minute—i.e., 28 beats above the basal, a rise of twenty-four per cent. The window was then closed. Nine minutes thereafter the rate had fallen from 146 to 134. It continued to fall until, sixteen minutes after the window was closed, it again reached the basal level. I would stress the fact that the animal throughout the test was resting quietly so that the reaction was not at all complicated by adrenal secretion stimulated through excitement or bodily movements. That the sympathetic system was active, however, was made evident by the erection of the hairs four minutes after the window was opened and by continuance of this state until about four minutes after the window was closed, when the animal was covered with a blanket.

In figure 28 is shown by a series of graphs the increase of heart rate above a basal level when animals with adrenals intact were exposed to cold air in the manner just described. The period of exposure is denoted by a thickening of the base line. The zig-zag or V-shaped marks represent shivering. Observe especially that the shivering is not a necessary condition for the faster heart rate, because the rate was increased long before the shivering began. There is clear proof that the faster pulse is due to adrenal secretion. In cases 46 and 49, with adrenal secretion excluded, the dash-lines show the changes of the heart rate under the usual conditions of exposure to cold. Note that the primary effect was a decrease of the rate instead of an increase.

The defect of the method just described lies in its limitation to periods of cold weather. In order to have a method which could be applied at any time we devised the procedure of introducing into the stomach a known amount of ice water. Of course, ice water can be given at any time and

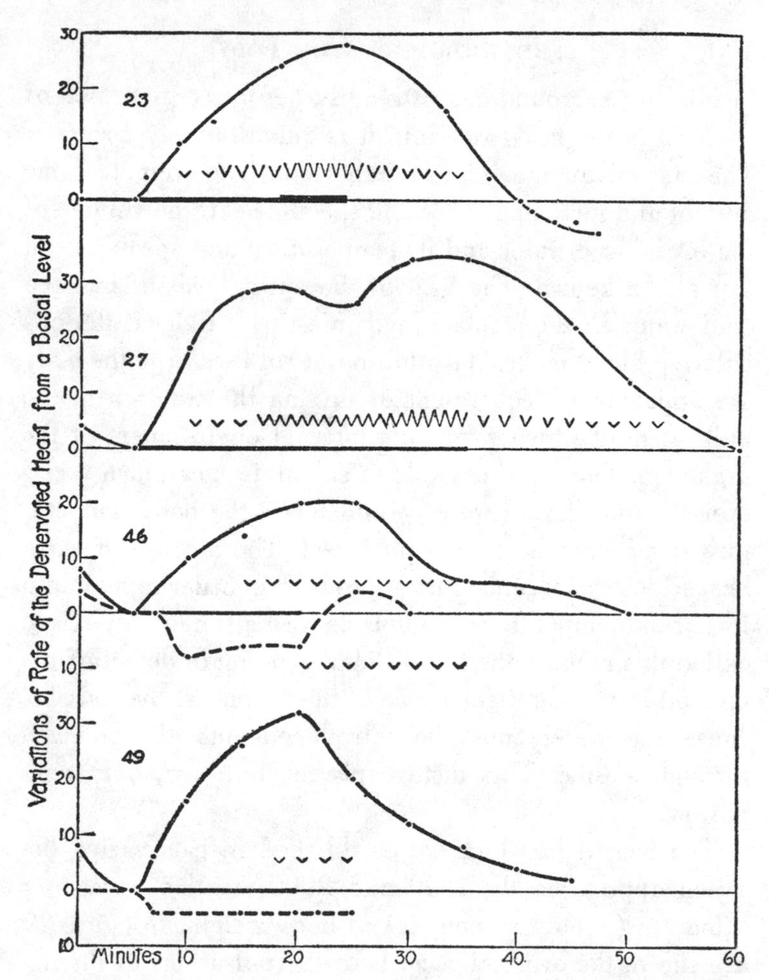

Fig. 28. Increases of rate of the denervated heart above a basal level when animals with adrenals intact (nos. 23, 27, 46 and 49) were exposed to cold air. The period of exposure is denoted by a thickening of the base line; the double thickening in the case of no. 23 marks the time of a cold draught through an open door. Shivering is indicated by separate v's when it was intermittent, and by connected v's when it was continuous. The size of the v's marks roughly the intensity. The dash lines in cases 46 and 49 show the changes of heart rate under similar conditions of exposure after exclusion of medulliadrenal secretion.

in familiar surroundings. It readily permits registration of changes in the heart rate and it is quantitatively accurate. The last advantage is most important. The weight of the animal and its temperature and specific heat, the volume of the introduced water and its temperature and specific heat, can all be known. The heat of the animal passes into the cold water in the stomach and intestines. Indeed, the circulating blood makes the addition of cold water to the body the approximate equivalent of mixing the water with the bulk of fluid which forms the internal environment of the organism. Thus it is possible to calculate how much a certain amount of cold water would lower the body temperature if no extra heat were produced. The amount of extra heat which the organism must produce in order to maintain its normal temperature can thus be well estimated. We have called this amount the "heat debt." The main defect of the method is a slight disturbance of the animal at the start because the water must be introduced into the stomach through a tube. This disturbance is, however, only temporary.

The heat debt which we established in cats varied between approximately 1500 and 2000 small calories per kilogram (about two pounds) of body weight. In figure 29 are shown the original records of the pulsations of the denervated heart when cat 33 was given water at 1.0° C. (33.8° F.) in an amount which established a heat debt of 1850 small calories. Observe that there was an initial increase in rate of 42 beats, due, no doubt, to the disturbance of giving the water. This soon dropped 8 beats, and thereafter for more than a half-hour the rate continued at a high level. Indeed, a full hour after the heat debt was estab-

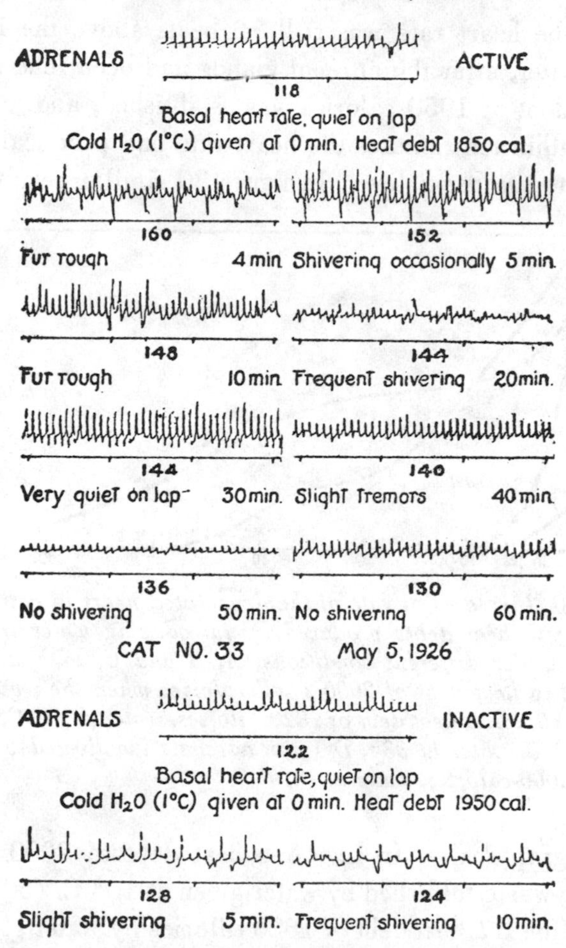

Fig. 29. Original records showing increases of rate of the denervated heart when cat 33 with active adrenals was given, April 16, 120 cc. of water at 1° (heat debt, 1850 small calories), and when, with adrenals inactivated, it was given, May 5, 116 cc. of water at 1° (heat debt, 1950 calories). Below each record are stated the prevailing conditions and the interval since the water was given. Time in five-second intervals.

lished the heart rate was still 12 beats above the initial level. Later, after the adrenal glands had been inactivated a heat debt of 1950 calories was established, and, as you note, within ten minutes the heart rate had practically returned to the basal level. In figure 30 similar results are

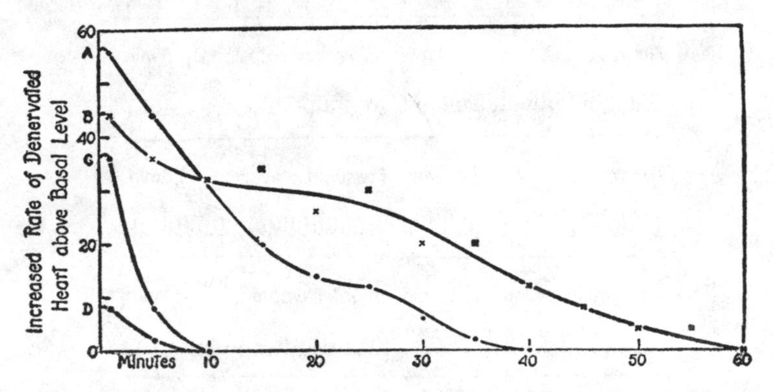

Fig. 30. Increases of rate of the denervated heart in cat 27, responding to heat debts produced by introducing water into the stomach under different conditions. A, B and C, with adrenals active. A, a heat debt of 2000 small calories when the water was given at 10°. B, a heat debt of 1820 calories, water at 1°. C, a heat debt of 250, water at 33°. D (with adrenals inactivated), a heat debt of 2000 calories, water at 1°.

shown graphically. In line A a heat debt of 2000 small calories was established by water given at 10° C. (50° F.), and in line B a heat debt of 1820 calories by water given at 1.0° C. The colder water had a considerably longer effect. In line C a small heat debt of 250 calories was established by giving water at 33° C. (91.4° F.). The main effect here was that of excitement; within ten minutes all of the disturbance had subsided.

From these and many other similar observations we drew

the conclusion that conditions which would naturally cause a reduction of body temperature induce an increased discharge of adrenin into the circulating blood.

V

Various investigators have found that adrenin has a heat-producing effect. Boothby and Sandiford have demonstrated that a milligram of adrenin injected into a man augments the basal heat production by fifty large calories. Since adrenal secretion is increased when the organism is exposed to the danger of a too rapid loss of heat, and since adrenin secreted in natural amounts is capable of accelerating the oxidative processes, it should be possible to demonstrate the service which this physiological reaction performs for the organism. We attempted in two ways to test the value of the reaction. It seemed possible that the heat produced by the automatic muscular contractions in shivering would be relied upon to a greater degree if the adrenal glands were rendered inactive, and that increased metabolism could be demonstrated without the assistance of the shivering mechanism. Accordingly we studied the effects of a certain heat debt on shivering when the adrenal glands are active or inactive, and we observed in man the influence of a heat debt on the metabolic rate in the absence of shivering.

Let us consider first the effect of a heat debt on shivering when the adrenal glands are present or absent. If the heat debt is large, i.e., if it amounts to 1000 small calories or more per kilogram, with the water at 1.0° C., and if the room temperature is about 20° C., it is commonly met by two calorigenic agencies: by an increased output of adrenin and by shivering. In figure 28 the fact is demonstrated that shivering coincides with the period of greatest discharge of adrenin. I emphasized the point, however, that shivering is not a necessary condition for the discharge of adrenin; indeed, shivering may be wholly absent although the heart

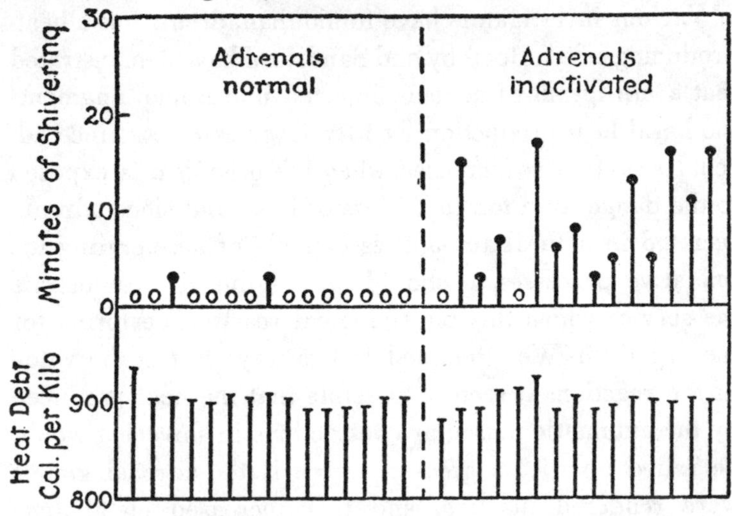

Fig. 31. Presence or absence of shivering when there was a heat debt of about 900 calories in a series of cats with normally innervated adrenal glands, and in another series with one adrenal gland removed and the other denervated.

rate is well accelerated. Now if the environmental temperature is about 20° C. and a heat debt of only 900 calories is to be paid, we found that shivering rarely occurs, or if it occurs it is of short duration. As shown in figure 31, in fifteen tests made in these circumstances shivering occurred in only two instances and lasted only three minutes. Note, on the other hand, what occurs if one adrenal has been removed and the other denervated. The same heat debt, established under the same conditions, resulted in shivering in all but two instances and it lasted as long as fifteen, sixteen and seventeen minutes. Thus when the heat-producing service of the adrenal medulla is lacking the shivering mechanism is resorted to.

The other observations, to test the effect of a heat debt on the metabolic rate in the absence of shivering, were made on human beings. We established a heat debt which averaged 449 small calories per kilogram. In twenty-two observations on eleven human subjects there was an average maximal increase in metabolism of a little more than 16 per cent, with variations above that average ranging as high as 38 per cent. These increases in the metabolic rate were not accompanied by shivering. The reader might suppose that the effect was due to the disturbance of taking the cold water and ice which established the heat debt. The highest point of the increase, however, averaged twenty-three minutes after the water and ice were taken and therefore occurred too long after the ingestion to be due to that. Furthermore, when warm water, equivalent in amount to the cold water, was drunk, the average increase in metabolism was only 3.1 per cent, and since the time of maximal increase occurred regularly in the first seven minutes of the experiment, it must have resulted chiefly from the disturbance of taking the water. This fact is brought out clearly in figure 32. A heat debt of 46 small calories per kilogram, due to the taking of 520 cubic centimeters of water at 30° C. (86° F.), caused in J. L. H. an immediate increase of 4 per cent in the metabolic rate. Later a heat debt of 409 calories per kilogram, due to the drinking of 354 cubic

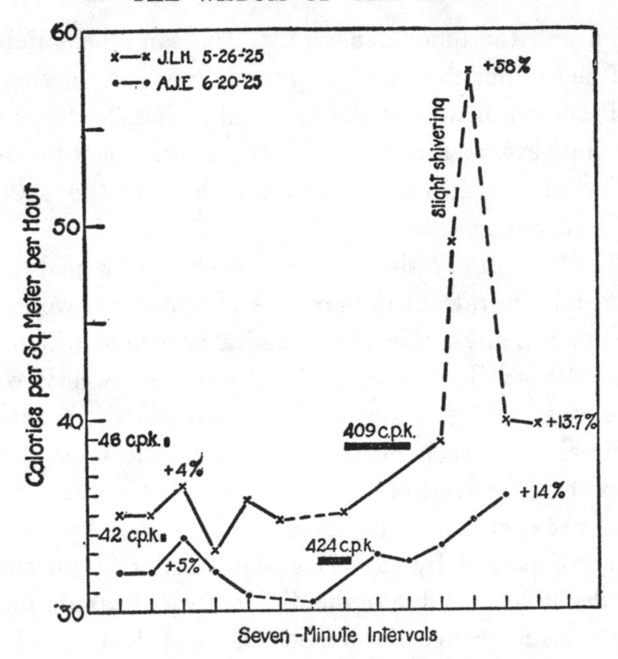

Fig. 32. Changes in metabolism produced by taking warm water and later by taking cold water and ice. A heat debt of 46 small calories (per k.) due to taking 520 cc. of water at 30° caused in J. L. H. an immediate increase of 4 per cent in the metabolic rate; a heat debt of 409 calories from taking 354 cc. of water at 1° and 130 g. of ice caused an increase of 13.7 per cent. Slight shivering was attended by a sudden and temporary increase of 58 per cent. A heat debt of 42 calories from 420 cc. of water at 30.2° caused in A. J. E. an immediate increase of 5 per cent in the metabolic rate; a heat debt of 424 calories, from 260 cc. of water at 1° and 139 g. of ice, caused a gradual increase of 14 per cent, without shivering. Each point breaking the line represents a metabolism record.

centimeters of water at 1.0° C. and swallowing 130 grams of ice, caused an increase of metabolic rate of 13.7 per cent. Note that a brief period of shivering was attended by a sudden and temporary increase of 58 per cent in the rate.

The observations on A. J. E. are quite free from the criticism that shivering might have affected the final result. In that case a heat debt of 42 calories from 420 cubic centimeters of water at 30.2° C. (86.4° F.) caused an immediate increase of 5 per cent in the metabolic rate. Later the taking of 260 cubic centimeters of water at 1.0° C. and 139 grams of ice established a heat debt of 424 calories. Observe that this caused a gradual increase in metabolism, amounting to 14 per cent, without any shivering whatever.

The foregoing experiments have shown that the same conditions that increase adrenal secretion in lower animals increase the metabolism in man and that they may do this without the accompaniment of shivering. It seems reasonable to conclude, therefore, that a disturbing heat loss evokes activity of the adrenal medulla in man as it does in the lower animals and that the extra output of adrenin in both organisms has the same effect of accelerating combustion.

It may be that the thyroid gland as well as the adrenal is involved in the processes of temperature regulation. We have seen that when it is overactive in disease, heat production in the body is greatly increased. Perhaps the thyroid is stimulated to action by cold and coöperates with the adrenal glands, but in a less quickly responsive manner, to accelerate oxidation. At least there are some observations which suggest that possibility. When cattle on our Western plains are first exposed to cold weather in early winter an enlargement of the thyroid gland has been noted as a characteristic change in them. And Loeb has reported that when part of the gland has been surgically removed the remnant grows more actively if the animal is placed in cold surroundings than if it is kept warm. But this is only suggestive evidence. Here again, more information is needed before we can draw definite conclusions.

As we survey the arrangements which check a shift of body temperature in one direction or the other it is interesting to observe that there are successive defenses which are set up against the shift. If dilation of the skin vessels does not stop the rise of body temperature, sweating and even panting supervene. If conservation of heat by constriction of the skin vessels does not prevent a fall of temperature, there is a chemical stimulation of more rapid burning in the body by means of secreted adrenin, and if that in turn is not adequate to protect the internal environment from cooling, greater heat production by shivering is resorted to. It is noteworthy that in all these functions except shivering the sympathico-adrenal mechanism is at work. As investigations by Dworkin in the Harvard Physiological Laboratory have shown, shivering itself has its most complete expression when that part of the brain, the diencephalon (see fig. 33, p. 233), which is the coördinating center for the sympathetic system, is intact.

We must recognize that among civilized people the physiological devices for the maintenance of constant temperature may have little opportunity to function. In wintry weather we spend our days in heated houses and offices and travel about in heated cars. Encased in warm clothing we carry with us everywhere a temperate climate. Thus only few occasions arise which demand either the conservation of the heat always being produced by the organism, or the development of extra heat by bodily activity. And in summer, likewise, mechanically operated fans, cold drinks, ice cream and refrigerated rooms lessen the use of the natural arrangements for keeping cool. It is not impossible that we lose important protective advantages by failing to exercise these physiological mechanisms, which were developed through myriads of generations of our less favored ancestors. The man who daily takes a cold bath and works until he sweats may be keeping "fit," because he is not permitting a very valuable part of his bodily organization to become weakened and inefficient by disuse.

VI

The delicate control of body temperature indicates that somewhere in the organism a sensitive thermostat exists which regulates the operations which we have been considering. Experiments on rabbits have shown that this part of the regulatory apparatus is located in the base of the brain, in the diencephalon (see fig. 33). The cerebral hemispheres and other parts anterior to that region can be removed, as Isenschmid has shown, and although the surrounding temperature may vary from 10° C. (50° F.) to 28° C. (82.4° F.), the normal temperature of the animal is preserved. If now the diencephalon is separated from the rest of the body, heat regulation is lost; the reactions to temperature changes in the environment are those of a cold-blooded animal. It is interesting to note that in the diencephalon are the central stations for the secretion of sweat, for shivering and probably also for panting—in short, for the automatic reactions which govern the production and loss of heat.

The thermostat in the diencephalon may be affected in

two ways, either by the temperature of the blood going to it or by nerve impulses from the surface of the body. Warming the blood in one of the large arteries in the neck that distributes to the brain, will cause dilation of the blood vessels of the skin and sweating. On the other hand, cooling that part of the circulation will cause shivering. And Leonard Hill found that sweating in a hot room could be stopped by placing his hand in cold water but not if the circulation through the arm was checked. Persistence of the local sensation of cold proved that nerve connections had not been disturbed. It is clear, therefore, that the flowing blood itself may affect directly the regulatory center. The evidence for reflex nervous influence on this center is also clear. A sudden dash of cold water on the skin causes not a fall of temperature but a rise, due to reflex constriction of the surface vessels and consequent interference with the normal loss of heat. Furthermore, the observation has been reported that if a person takes a bath in water at 29° C. (84.2° F.), he feels cold, shivers, and by the reaction keeps constant his body temperature. If he takes a bath at the same temperature in carbonated water, however, he does not feel cold, there is no reaction, and the body temperature falls. Although we have this evidence of a double control of the responses of the thermostat in the brain, the actual mode of control—for example, the influence on it of the nerves which give us sensations of heat and cold—is not yet well understood.

In spite of gross interference by civilized man with the physiological mechanisms for homeostasis of body temperature we know that these mechanisms exist and are always ready for action. If conditions are such that there is a tendency to tip the organism in one direction, a series of processes are at once set at work which oppose that tendency. And if an opposite tendency develops, another series of processes promptly oppose it. Thus quite automatically the remarkable uniformity of the temperature of the internal environment is preserved, in opposition to both internal and external disturbing conditions.

References

Aub, Bright and Forman. Am. Journ. Physiol., 1922, lxi, 349.
Benedict and Carpenter. Publication No. 261, Carnegie Institution, Washington.
Britton. Quart. Journ. Exper. Physiol., 1928, xiii, 55.
Cannon, Querido, Britton and Bright. Am. Journ. Physiol., 1927, lxxix, 466.
Cannon, Newton, Bright, Menkin and Moore. Ibid., 1929, lxxxix, 84.
Cramer. Report, Imper. Cancer Research Fund, London, 1919, 1.
Dworkin. Am. Journ. Physiol., 1930, xciii, 227.
Hartman, McCordock and Loder. Ibid., 1923, lxiv, 1; cf. also lxv, 612.
Hill, Journ. Physiol., 1921, liv, p. cxxxvi.
Isenschmid. Hanbk. d. norm. u. path. Physiol., Berlin, 1926, xvii, 56.
Loeb. Journ. Med. Res., 1920, xlvii, 77.
Lusk. Journ. Physiol., 1924, lxix, 213.
McIver and Bright. Am. Journ. Physiol., 1924, lxviii, 622.

XIII

THE AGING OF HOMEOSTATIC MECHANISMS

THUS far we have considered the organism as if it were always ready to meet emergencies and as if the mechanisms for homeostasis were invariably efficient. In the years from birth to old age, however, there are changes which can be observed, both in the constancy of the fluid matrix and in the ability of the physiological devices which ordinarily maintain the stable state.

I

Mechanisms for preserving homeostasis are obviously not required by the developing mammal before birth. Vigorous exertion does not occur, and the immediate environment of the fetus *in utero* is kept constant by the homeostatic agencies in the mother's body. Furthermore, the fetus has external fluid contacts which do not differ greatly from its own internal fluids. At birth, however, the infant is suddenly introduced into gaseous, cold, rough surroundings; a gasp draws into the lungs a thin, dry air, and thereafter every inward breath may cool the body by absorbing heat and by favoring evaporation, and every outward breath carries away water vapor and carbon dioxide.

To preserve a steady state of the fluid matrix in this

strange and foreign world requires physiological operations which have not previously been exercised. Time is required to render them efficient. If new-born babies, therefore, are exposed to even a moderate degree of cold they suffer a sharp drop of body temperature. And, as Schretter and Nevinny have noted, the blood-sugar concentration of infants oscillates from day to day and from hour to hour much more than that of the adult. It is quite possible that if other states were carefully examined, which in later life are kept relatively constant, they would show in early childhood a period of variability preceding the remarkable steadiness which is typically developed.

As organisms grow older they manifest an increasing tendency to be indolent. This is a quite general phenomenon. One has only to compare the frolics of a puppy or a young dog with the slow movements and the somnolence of an old dog, in order to have a striking illustration. Our own motions, also, become gradually fewer and slower as the years of life pass onward from the third or fourth decade. Associated with this indolence and retardation are interesting correlated changes in homeostatic mechanisms. We shall consider these changes in relation to three homeostatic conditions—the uniformities of temperature, blood sugar and the acid-base balance.

II

Temperature homeostasis, as previously shown, is the result of regulating heat production and heat loss. A fact of primary importance is that the internal temperature of elderly persons is usually kept within the normal range. Observations by Martsinkovski and Zhorova, made on 185 individuals ranging in age from 60 to 100, revealed no alteration of body temperature as the years increase. Though it is thus maintained, are the factors which maintain it undisturbed in the course of growing old? This is the central question.

Fundamentally, heat is produced by the processes of chemical change taking place in the organism. The common measure of these processes is the basal metabolic rate (see p. 180). Various observers, since the time of Magnus-Levy and Falk in 1899, have testified that the speed of metabolism becomes gradually less with advancing years. The greatest accumulation of pertinent data on this point was gathered by Boothby and his collaborators at the Mayo Clinic. They reported tests on 639 male and 828 female human subjects. The average figure for men at age 20 was 41.6 calories per square meter per hour. At age 40 it had fallen to 38.3 and at age 60 to 35.7—a drop of about 14.5 per cent from age 20. Studies by Matson and Hitchcock and also by Benedict, on men whose ages ranged from 74 to about 90, revealed an average basal metabolic rate of about 30 calories per square meter per hour. This represents a fall of about 28 per cent from age 20. In round numbers we may expect that in late senescence the basal heat-producing processes in the body will be reduced about 25 per cent, as compared with the years of early manhood.

The reasons for the reduced rate of heat production as we grow older are probably various. Pertinent among them are, first, indications of partial involution of the thyroid gland. In the thyroid gland of the aged, according to Dogliotti and Nizzi, clusters of small follicles are found with little colloid in them; other clusters of follicles are much distended with colloid. Also, there is a gradual increase of fibrous tissue, with lymphoid infiltration. The great importance of the thyroid as an agent maintaining normal heat production renders these observations significant. Another reason for a lowered metabolic rate may be muscular weakness and associated reduction of muscular vigor. Quetelet tested the ability of persons of various ages to lift heavy weights. He found that this ability was approximately 40 per cent less at age 60 than it was between ages 25 and 35. Whether there is an associated reduction in muscular tone and therefore a reduced output of heat is not definitely known.

The explanation of a lessened heat production as one grows older is admittedly unsatisfactory. There remains, however, the fact that it is a limiting condition in the homeostasis of body temperature. The gradual decrease in the rate of combustion in old age still leaves enough heat produced to maintain normal temperature in ordinary circumstances. On exposure to cold, however, the slower burning in the body must be compensated for by putting on more clothing than is required in youth or middle age and by seeking a place near the stove or an open fire, or some other warm place.

Besides a lessened ability to adapt to cold there is often in the aged a lessened ability to adapt to external heat. It has been pointed out that this deficiency is more marked in elderly persons who are fat than in elderly persons who are lean. Sweating and vascular dilation may both be defective because the skin commonly undergoes atrophy in old age. A partial disappearance of dermal capillaries may occur, with attendant degenerative changes in collagen and in elastic tissue. Furthermore, there is likely to be partial degeneration of sweat and sebaceous glands so that the skin becomes dry and rough in the later years of life. Also, with advancing age the internal layer (the intima) of the arterioles may thicken, and there may be a development of fibrous tissue in the muscular media; in consequence of these changes vasodilation of the arterioles may be reduced, and therefore the possibilities of heat loss by vasodilation are correspondingly reduced.

Pickering has studied the maximal heat elimination from the hand when it was exposed to a standard amount of water at standard temperature. He found that the output of heat in calories per minute per unit volume of hand was 33 per cent lower at 70 years of age than at 25. These observations are quite consistent with the changes in skin and blood vessels which we have just surveyed.

Concordant with this evidence of lessened ability to get rid of body heat, or to adjust to increased temperature, is the evidence of an increased death rate from "heat stroke" in the later decades. The rate rises slightly after 60, and then goes up with striking rapidity. The figures for Massachusetts between 1900 and 1930 showed a death rate from heat stroke of 8 per 100,000 between the ages of 70 and 79; 20 between 80 and 89; and 80 between 90 and 100. Similar figures have been reported for the states of New York and Pennsylvania.

In reviewing the facts thus far considered in relation to the homeostatic devices concerned with the regulation of body temperature, note that the powers of adjustment to high and low external temperatures are greatly restricted as one grows older.

III

We turn now to a consideration of the homeostasis of blood sugar. This is achieved, as noted previously, by storage in the form of glycogen in times of plenty, by overflow through the kidneys when there is excess of glucose in the blood, and by release from storage in times of need. The efficacy of the mechanisms for storage and for use of glucose in the body is measured by the so-called "glucose-tolerance test." After a standard solution of glucose is taken on an empty stomach the concentration of sugar in the blood is determined at intervals of a half-hour, usually for about two hours. Normally there is a sharp rise in the concentration during the first half-hour to 160 or 170 mgm. per cent and thereafter a gradual return to the previous level in about two hours. When the mechanisms are defective the resultant curve, indicating conditions in the blood, is higher and longer than the normal.

Various observers have noted that in the later decades of life the test reveals a higher and more prolonged rise in the glycemic concentration than is characteristic of the earlier decades. The largest number of cases have been reported by John, who made the test on 192 children and 1500 adults. He found that the "diabetic" type of curve—a high and prolonged hyperglycemia typical of diabetes—was found in only 10 per cent of individuals between 30 and 40 years of age, and that there was a gradual increase to 50 per cent between ages 60 and 70. Of children, 80 per cent had normal curves for glucose tolerance, whereas only 62 per cent of adults were in this class.

These data indicate that with advancing years the human

organism is prone to an impaired ability to use and store glucose at the rate characteristic of youth and early adulthood. Again we note that the homeostatic mechanisms become limited as we grow older. Unfortunately only partial knowledge of the regulation of the glycemic level prevents us from stating definitely what factors may be defective.

IV

The chief homeostatic agents narrowly maintaining the slightly alkaline reaction of the blood are, as already explained, the lungs, the heart and the blood vessels. A large production of lactic acid is associated with strenuous muscular activity—for example, a vigorous struggle (see p. 147). Unless this non-volatile acid is promptly burned to volatile carbonic acid, it will overwhelm the alkaline buffers in the blood; these are the first defenses of the organism against the development of an acid reaction which, even when slight, is highly dangerous. In reviewing previous evidence we note that the burning of lactic acid requires an abundance of oxygen in the active tissues. This is provided by an increased depth and rate of respiration, by a rise of general arterial pressure with dilation of the blood vessels in the active parts, and by a larger output of blood from the heart per beat, and also per minute because of a faster rate. The consequence of these circulatory changes is a greatly increased use of the carriers of the respiratory gases between the tissues and the lungs—oxygen to the tissues and carbon dioxide to the lungs. The questions, then, which are raised, relate to the changes, as age advances, in the homeostatic mechanisms concerned with the avoidance of accumulation of acid in the blood.

The first change which we may note is a lessening of the capacity for ventilation of the lungs. As long ago as 1846, Hutchinson reported on the maximal to-and-fro movement of respired air (i.e., the vital capacity) in 1775 healthy cases. He found that there was a gradual diminution of the vital capacity from a maximum between ages 30 and 35, until in the period between 60 and 65 it was only 80 per cent of the earlier figure. The normal vital capacity of young adults, as determined in a large number of cases, is approximately 3750 cubic centimeters for males and 2600 for females. In an examination of 110 men between 60 and 94, and 71 women with approximately the same range of ages, Levy confirmed Hutchinson's figures for the period 60–65 and discovered that in the late 80's the average vital capacity of males is reduced to 2350, a fall of 37 per cent, and in females to 1460, a fall of 43 per cent from the averages for young adults. It might be supposed that the vital capacity, as measuring a maximal effort, might fall without a corresponding fall in the ordinary physiological capacity of the lungs. It appears to be true, however, that as the vital capacity is reduced in old age the pulmonary ventilation in maximal muscular work is similarly reduced. Studies by Dill and Robinson have disclosed that in elderly men the respiratory movements in supreme exertion are reduced to a degree corresponding to the diminished vital capacity, or to even a greater degree.

The lessened mobility of the thoracic wall may be attributed to a weakness of the intercostal and other respiratory muscles which may result from lack of exercise or the indolent habits of the aged. It may be attributed also, and perhaps more reasonably, to a stiffening of the attachments of the ribs. Determinations of the calcium content of the costal cartilages by Bürger and Schlomka have shown that whereas it amounts to 125 milligrams per cent of dry substance in youth, it rises to 617 milligrams per cent in the fifth decade and to approximately 1400 milligrams per cent in the seventh decade. The accumulation of lime salts in the flexible attachments of the ribs may result in such rigidity that a deep intake and vigorous output of air, possible in earlier years, become impossible as one grows older.

Vascular changes occurring in senescence likewise induce limitations. First, the arterial pressure commonly rises as men and women pass into and beyond the decade of 30 to 40 years. In a statistical study of 4000 cases Saller found the average systolic arterial pressure in men between 21 and 47 years of age was 144 millimeters of mercury, and between ages 68 and 89 it was 186. Similar changes occurred in women in the same age groups, but in general the figures were higher than for men. Although these statistical data are more extreme than have been reported by other observers, they agree with other data in showing the tendency of the systolic arterial pressure to rise as the decades pass.

Another change which is probably associated with the increased blood pressure is a diminishing elasticity of arteries with advancing years. The elasticity has been estimated by determining the velocity of the pulse wave. Bramwell, Hill and McSwiney measured the velocity of the wave in 74 individuals between 5 and 84 years of age. The elasticity was expressed as 47 per cent at 5 years; it was found to be reduced to 17 per cent at 80 years. Hallock, using a more delicate method for registering the pulse wave and studying 400 persons with ages ranging from 5 to 65 years, found a close correspondence with the results reported by the earlier observers up to age 45, and thereafter a more rapid loss of elasticity than in the smaller group. Obviously, if the arteries are more rigid everywhere in the body and not merely in the superficial samples taken for examination, they could expand less in the active muscles of the old than in the young.

There is evidence that capillaries as well as arteries may have impaired functions as the years pass. In a histological examination of muscles Buccianti and Luria have found that there is a laying down of interstitial colloid in old persons and a thickening of the elastic tissue which surrounds the muscle fibers. In this interfibrillar region lie the capillaries. If this extra material is interposed between them and the muscle cells it is obvious that the capillaries, even should they dilate when the muscle becomes active, could not perform their function satisfactorily, for the diffusion of the respiratory gases—especially oxygen, which has a relatively slow diffusion rate—would meet with obstruction.

The third homeostatic factor which has been mentioned is the heart. Observations on cardiac performance in senescence have shown that that organ does not meet the demands in later so well as in earlier decades. Dill and Robinson have made records of the cardiac rate of ninety-one boys and men ranging in age from 6 years to nearly 70. These subjects were required to run on a treadmill having a slope of 8.6 per cent. The speed of the treadmill varied according to the subjects' ability to run. It was sufficiently great, however, to render almost all the subjects exhausted in three to five minutes. The mean of the maximal heart rate of nine boys having an average age of about 14 years was 196 beats per minute. In seven men with an average age of 63 years it was 165 beats per minute. The intervening ages showed a continuously downward trend in the ability of the heart to keep the faster pace as the years increase.

It seems probable that in the elderly the heart is not only beating at a less rapid rate but also with a reduced efficiency. The "hypodynamic heart" of the untrained person functions differently from the vigorous heart of the athlete. When the well-trained individual engages in vigorous exertion the heart not only beats somewhat more rapidly but also empties itself more completely of the blood delivered to it from the veins (see p. 157). The "hypodynamic heart" of the untrained person, on the other hand, meets the situation differently. It dilates and thereby has the physical advantage of greater length of muscle fibers, but it does not empty itself effectively, according to Wiggers; the increased output per minute, therefore, depends largely on the more rapid beat. This suggested difference between the young and old requires, however, further evidence before it can be accepted as reliable.

V

In surveying the homeostatic mechanisms concerned with protection against the development of an acid reaction in the blood—the lungs, the blood vessels and the heart—we note that as life goes to its later stages there is a gradual narrowing of the capacity to adapt the needs of the organism to the special requirements presented in muscular effort. These mechanisms work satisfactorily in a routine, quiet existence, but stress, even relatively slight stress, may encroach sharply on the limitations.

What are the results of these limited capacities of the lungs, blood vessels and heart? Mori subjected youths and men, ranging in age from 17 to 57, to standard work for ten minutes on a bicycle ergometer. In four youths of the second decade the alkali reserve (see p. 46) of the blood was lessened to a degree indicated by a reduction of 4 volumes per cent of carbon dioxide. In thirteen men of the fifth decade the reduction was 12 volumes per cent of carbon dioxide. It is probable that in these cases the lessened alkali reserve was due to an increase of lactic acid coming from the laboring muscles. That inference is supported by observations made by Dill and Robinson. Their subjects walked on a treadmill having a slope of 8.6 per cent. A rate of walking 3.5 miles per hour increased the metabolism until it was seven times the basal. Examination of the blood in the subjects showed that the lactic acid which was present was more than three times as much at age 60 as it was at age 20. Fortunately, in these studies we have definite facts on which to base the explanation of the greater appearance of lactic acid in the older subjects. It should be emphasized that all subjects were performing fairly hard work. A measurement of the oxygen carried from the lungs per minute in these individuals was gradually less with advancing years. At age 17 it was 53 cubic centimeters per kilogram per minute, at age 35 it was 43 cubic centimeters, and at age 63 it was 35—a drop of 34 per cent. These are figures which have their reasonable explanation in the lessened ability of lungs, blood vessels and heart to meet the demands of vigorous effort as one grows older.

Concordant with the foregoing data is the evidence presented by records in competitive sports. The best times for the 100-yard dash have been made by youths or young men; the present record was first made by Wykoff at age 21. The records from one to five miles are held by men with ages ranging between 23 and 27 years. The record for ten miles was made by Nurmi when he was 31 years old. DeMar, after running Marathon races from the time he was 22 years old until he was 50, made his best showing between the ages 36 and 42. It appears that, as speed alone becomes less important for running and judgment and endurance become more important, the records are held by older men.

Reports of the performance of baseball players and tennis champions likewise furnish evidence that the ability to mobilize quickly the bodily forces is gradually lost after a peak at the period between 30 and 35 years of age. Experts have testified that after age 35 the professional players of baseball begin to "slow up," their "legs fail," they lose the speed they had earlier. There are few stars in sport after age 40.

In a final survey of the facts which we have considered we note that in temperature regulation, in the storage and use of sugar in the body, and in the maintenance of the acid-base balance of the blood the homeostatic mechanisms, when subjected to stress, are revealed to be more and more limited in their ability to preserve uniformity of the internal environment as life advances into the last decades.

References

Benedict. New England Journ. Med., 1935, ccxii, 1111.
Boothby, Berkson and Dunn. Am. Journ. Physiol., 1936, cxvi, 468.
Bramwell, Hill and McSwiney. Heart, 1923, x, 233.
Buccianti and Luria. Arch. Ital. di Anat. e di Embriol., 1934, lxxxiii, 110.
Bürger and Schlomka. Zeitschr. f. d. ges. exper. Med., 1927, lv, 287.
Dill and Robinson. Personal communication, 1938.
Dogliotti and Nizzi. Endocrinol., 1935, xix, 289.
Hallock. Arch. Int. Med., 1934, liv, 770.
Hutchinson. Med.-Chir. Trans., Roy. Soc. Med. and Chir., London, 1846, xxix, 169.
John. Endocrinol., 1934, xviii, 75.
Levy. Zentralbl. f. inn. Med., 1933, liv, 417.
Magnus-Levy and Falk. Arch. f. Physiol., 1899, Suppl. Bd., 314.
Martsinkovski and Zhorova. Acta med. Skand., 1936, xc, 582.
Matson and Hitchcock. Am. Journ. Physiol., 1934, c. 329.
Mori. Jap. Journ. Med. Sci., Biophysics, 1936, iii, 309.
Pickering. Clin. Sci., 1936, ii, 209.
Quetelet. Quoted by Rubner. Deutsch. med. Wochenschr., 1928, liv, 1750.
Saller. Zeitschr. f. d. ges. exper. Med., 1928, lviii, 683.
Schretter and Nevinny. Arch. f. Gynakol., 1930, cxl, 465.
Wiggers. The Circulation in Health and Disease. Philadelphia, 1923, p. 411.

XIV

NATURAL DEFENSES OF THE ORGANISM

IN A sense the preservation of constancy in the fluid matrix of the body may properly be regarded as a defense against unfavorable conditions which would arise if the constancy were not preserved. And there is little doubt that the mechanisms of homeostasis belong to the general category of protective functions. But special reactions of the organism, nicely adopted to dangers which might arise from the presence or attack of injurious external agents, warrant at least a brief survey of some types of these defensive arrangements. Let us consider first the protective reflexes.

I

If a foreign body enters one of the nostrils it produces a peculiar tickling sensation which is followed by a sneeze. The mucous membrane of the nose has been stimulated, an afferent impulse has passed to the base of the brain by way of a sensory nerve, and, without the requirement of an act of will, motor impulses are discharged into the muscles of respiration. First a deep inspiration is taken through the mouth, and then the inspired air is violently driven forth through the nose and mouth. As the blast rushes out through the nasal passages it is likely to carry with it the irritating object. Thus by a powerful reflex of expulsion the lower respiratory passages are protected from the irritant affecting the upper passages.

Let us suppose, however, that the object has slipped past the defenses in the nose, or has been breathed in through the mouth, or has "gone down the wrong way" on being swallowed. It finds in the larynx an almost impassable barrier. Immediately the inspiratory movement, which might carry into the lungs the offensive agent, is sharply and completely stopped, due to afferent impulses, borne to the brain by the superior laryngeal nerve, that have an inhibitory effect on the respiratory center. But this arrest is only the first phase of the reflex. At once a vigorous expiration is provoked—a cough. The foreign body is thereby swept upward out of the larynx. It is a highly interesting fact that stimulation of the superior laryngeal nerve not only causes a reflex cough, but also a reflex act of swallowing, so that anything forced out of the windpipe and larynx would be quickly carried down the esophagus. Of course, such reflex effects are induced also by objects or particles which enter the larynx from below, just as well as by those which enter it from above. It is thus that we clear out the secretions from the windpipe and bronchial tubes when we have a "cold" in the chest. The primary importance of these reflexes is proved by the almost certain development of pneumonia when the sensitivity of the larynx is abolished by section of its nerves, and when the cough reflex, therefore, is no longer on guard.

Another group of protective reflexes is found in connection with the ingestion of food. As Pavlov's experiments on dogs with an artificial opening of one of the salivary ducts have shown, any substance, placed in the mouth, that is irritant (e.g., acid) or likely to interfere with free movements of the parts on one another, promptly evokes a copious outpouring of saliva. The result is that the substance is quickly diluted so as to be less irritant, or, if disturbing to movements, it is washed away.

When a mass too large to be swallowed is thrust to the back of the mouth a spasm of the pharynx occurs, which may be followed by gagging and expulsion of the offending bolus.

Again, if an irritant substance or a substance harmful to the organism is propelled down into the stomach, it may be quickly ejected. Based on that fact is the common domestic method of giving mustard in a drink in order to make the stomach empty itself. If the substance is not immediately distressing, it occasions a remarkable series of events. First, in accompaniment with a feeling of nausea there is a free secretion of saliva, which is repeatedly swallowed. At the same time the troublesome stuff in the stomach stimulates a secretion of watery mucus from the glands in the gastric wall. These fluids dilute the gastric contents and thus render them less irritating and at the same time easier to discharge. As these processes go on and nausea increases, respiration becomes more rapid. X-ray studies reveal that the stomach is quite relaxed. Suddenly the diaphragm and the abdominal muscles contract simultaneously and thus increase greatly the pressure on the abdominal contents. Simultaneously, also, the narrow opening between the vocal cords (the glottis) closes, so that as the diaphragm descends it does not draw air into the lungs but lessens greatly the pressure in the thorax. The esophagus, which runs through this region of low pressure, dilates; the sphincter at the end of the esophagus opens; and the diluted gastric contents are forcibly ejaculated. In this manner the stomach itself, the rest of the gastro-intestinal tract, and the organism as a whole, are protected from the action of harmful material which has been swallowed.

The external surface of the eye is another region in which a special provision exists for protection against injury and invasion by foreign bodies. Besides being set in a secure bony socket, it has quick-acting covers in front which shut down whenever danger approaches. As soon as a foreign particle, however minute it may be, comes in contact with the eye, pain, effusion of tears and winking are instantly aroused. The pain demands immediate attention to the removal of the particle, the excessive secretion of tears helps to wash it away, and the winking brings a mechanical force to bear on the process. If these activities are unsuccessful, mucus may be produced to cover the object and protect the surface which moves over it. When a soluble irritant substance touches the eye the same reflexes occur, but the winking and the tears now coöperate to reduce the irritation by diluting the substance. In these ways the most wide-ranging of our sense organs, although it occupies a fairly prominent position, is shielded from harm.

II

Analogous to these protective arrangements of an immediate and reflex character are the more slowly developing processes which preserve the integrity of the organism. The appearance of a callous area on the skin, where a rough instrument repeatedly presses and rubs, illustrates that sort of protection; the callosity serves both as a cushion and as a shield. Again, if the skin is broken, and the wound happens to be clean, capillaries gradually appear in the clot which fills the gap in the surface, connective tissue cells develop as supporting structures around the fine vessels, the growing layer of the skin reaches out over this organized area, and, apart from a whitish scar, the region is finally as solid and durable as before.

The process of repair of internal organs is remarkable. Unlike the surface structures the viscera have not been exposed to repeated injury in the course of racial history; indeed, such injury as would rupture the intestines, for example, would be more likely to result in death than in repair. It is difficult to understand, therefore, how the viscera could be "trained," so to speak, in reparative processes or how there could be a basis for selection of such organisms as were able to develop these advantageous reactions. And yet, wherever the surgeon enters, whether the lungs, the liver or the brain, the disturbed tissues promptly become the seat of activities which result in healing and fresh organization.

Many years ago Murphy and I observed with the X-rays a curious phenomenon after the first part of the small intestine (the duodenum) had been cut across and sewed together again. Although peristaltic waves were passing routinely over the stomach, the sphincter at the outlet (the pyloric sphincter) held tight against them, and only after about five hours did it relax and permit the gastric contents to enter the injured gut. The interest here lies in the relation of the delay to the process of healing; according to surgical observations, about four hours are required after an intestinal suture for a plastic exudate to form and make a tight joint. It was after the proper time had elapsed for that process to come to completion, therefore, that the chyme from the stomach was allowed to advance. Similar results were obtained when the section and suture were made further along the alimentary canal. Much work must still be done before we understand fully the mode of action of the agencies which are engaged in the restoration of damaged parts.

Belonging to this group of slow adjustments to changed conditions are certain responses to the external environment that favor homeostasis. We have reviewed the complex of ready responses in the organism which appear when need for oxygen arises. But besides these there is a slow response, seen in persons who for days or weeks are subjected to low percentages of oxygen in the air. If a person starts living at a high altitude, e.g., 14,000 feet, the number of red corpuscles in a cubic millimeter of blood slowly increases from the normal (about 5,000,000) to 7,000,000 or 8,000,000. Of course, this rise in the percentage of oxygen carriers in the circulation, like the rise which follows quickly when the spleen contracts, results in a larger supply of available oxygen for use by the secluded tissues than would otherwise be possible. Since the diffusion-pressure of alveolar oxygen at high altitudes is so low that each corpuscle leaves the lungs incompletely loaded, the only immediate modes of compensating for that disadvantage are by employing a more rapid circulation of the blood to increase the number of trips made by the carriers from lungs to tissues, and by increasing as much as possible the number of carriers from the splenic reservoir, in the manner already described. But these are emergency measures. If the need continues it is met by greater activity of the blood-forming organs in the marrow of the long bones, and thus by a permanent enrichment of the blood in red corpuscles, until the individual returns to a lower altitude.

Quite analogous to this slow provision for homeostasis in the oxygen supply is the slow reaction of the organism to cold. We have reviewed the series of quick defenses which are resorted to whenever there is danger of a lowered body temperature: the constriction of surface vessels and the erection of hairs or feathers, to lessen heat loss; and the secretion of adrenin, and finally shivering, to increase heat production. But these again are emergency measures. In animals protected by hair, the hair grows much thicker and richer in the cold weather. It is in the winter that fur-bearing animals are trapped, and at the end of the winter that sheep are shorn. Unfortunately we know very little about the stimuli which cause these slow adjustments of the organism that protect against an unfavorable or harmful external environment.

III

Besides external physical conditions which may injure the body there are living beings—small and large—which may be dangerous. Residing in the mouth and intestines, in the nose, in the skin—indeed, on practically all the surfaces and in practically all the orifices of the body—are myriads of bacteria. In the main they are harmless, but among them are forms which are capable of causing serious inflammation or disease. The most potent immediate protection against these harmful forms is an intact and healthy body surface. Only when the skin is broken or cut, only when the mucous membrane is weakened or damaged, are the conditions favorable for the invasion of our bodies by these foreign enemies.

When disease-producing bacteria get a foothold they may do harm in one or more of three ways; they may attack locally, they may produce locally a poison which can enter the blood and injure the organism extensively, or they can enter the blood themselves and produce widespread damage.

The way in which our bodies meet a local attack is seen in the phenomena of a pimple or a boil. Perhaps because of failure to keep a part clean, or because of slight injury or some general weakened state, pus-producing bacteria, which are living close to the root of a hair in the skin, begin to multiply rapidly. As they do so they become irritating, probably by action of some material which they produce. As a consequence of the irritation, the capillaries of the region are injured, with consequent increase of the permeability of their walls. Thus plasma escapes from the blood stream into the tissue spaces where it coagulates in a fine fibrinous network. The local lymph vessels also are injured by the irritant and become closed by clots. The fibrinous net and the occluded lymphatics mechanically shut off the rest of the body from the inflamed area. Immediately after this barrier has been set up, leucocytes begin to gather in the region and ingest the imprisoned bacterial invaders. In this way the organism as a whole is protected at the expense of local injury. The accumulated material may cause swelling, with tingling or pain. The bacteria, living and dead, the leucocytes and the injured tissue, which becomes softened by a modified digestive process, form a whitish mass within the walled-off enclosure. The digestive process continues and gradually eats a way to the surface of the skin. When that is reached the mass, as "pus," is discharged. The local reaction in the skin has in this way saved the body as a whole from penetration by noxious living agents. The danger which is thus escaped is occasionally demonstrated by a careless person who pricks an inflamed spot before it is walled off and who as a result has a serious blood-poisoning.

The injury which bacteria produce by growing locally on some surface and giving off from there into the blood a toxic substance is well illustrated in the disease, diphtheria. The typical diphtheritic membrane produced by the bacteria may develop on almost any mucous membrane, but usually it is at the back of the throat. The superficial cells of the area are badly damaged, but the inflammatory reaction prevents the incursion of bacteria. The bacteria, however, as they grow in the membrane, produce a virulent toxin which, on diffusing into the blood stream and being distributed to all parts, has profoundly disturbing effects. They may include paralysis and death. To this circulating toxin the body responds by the development of a protective antitoxin which is capable of neutralizing the bacterial poison. It is interesting that this process can be duplicated in some of its aspects outside the body. If the bacilli of diphtheria, for example, are grown in bouillon they produce in it the characteristic toxin of diphtheria. On injection of the fluid into a lower animal (a guinea pig, for instance) typical toxic effects are manifest. But if the blood plasma or serum from an animal which has developed antitoxin is adequately mixed with the toxin in the bouillon, the bouillon loses its poisonous properties and can be injected without producing any disturbance. Repeated injections of diphtheria toxin in gradually increasing doses will cause an animal to develop antitoxin in large amounts and without the illness which attends the natural disease. It is this procedure which is employed in making antitoxin for the treatment of diphtheria in human beings. Persons who are both ignorant of biology and careless of the welfare of their fellow men have ridiculed this method of defense against a dangerous disease. They declare that the injection of antitoxin from a horse into a man is equivalent to injecting "filth." They do not realize that in developing diphtheria antitoxin in the horse, man is making use of one of the most extraordinary and beautiful processes of natural defense in the organic world, and that in applying that antitoxin to human beings afflicted with diphtheria he is merely strengthening the natural defense which the patients themselves set up. This procedure has led to our present control of the dread disease, diphtheria; and similar procedures, in other diseases of like character, are leading to similar control of them.

When the bacteria themselves, rather than a toxin which they produce, become the more important factor in the attack, the reaction in our organism is correspondingly directed against the invaders instead of their product. By producing an "antibody" the blood acquires the ability to cause the bacteria to clump into masses. When the change which induces clumping has been wrought, a natural substance in the blood plasma, the "complement," is often able to kill the bacteria, whereas, without the previous action of the antibody, it is impotent. Furthermore, the leucocytes and other cells capable of destroying bacteria by ingesting them, can engage in that defensive service much more effectively after the bacteria have been altered by the antibody than before. It is evident also that the destructiveness of these foraging cells depends on the number of bacteria which they take up and destroy, and that the number is sure to be greater, in a series of chance meetings, if the bacteria are in clumps or bundles than if they are single. The protective antibody developed in the blood acts, therefore, not only to sensitize the virulent bacteria, so that they are more easily destroyed by chemical means and by phagocytes, but also to arrange for their destruction on a large scale.

Just as the artificial introduction of toxin can be used to evoke an antitoxin in the blood, so, likewise, the injection of the bacteria of a disease which is caused especially by invasion can be used to evoke a specific antibody for such bacteria. In this procedure only dead bacteria or those which have been weakened by processes of cultivation are employed. It is thus that armies—and individual human beings—are protected against typhoid fever and its like.

IV

In the long history of the race bacteria have not been the only living foes of man, and in wild life, perhaps, they have not been the most important. There have been savage creatures, human and subhuman, watching with stealth and ready to attack without a moment's warning. And there has been, also, the necessity of fighting, for revenge, for safety and for prey. In that harsh school fear and anger have served as a preparation for action. Fear has become associated with the instinct to run, to escape; and anger or aggressive feeling, with the instinct to attack. These are fundamental emotions and instincts which have resulted from the experience of multitudes of generations in the fierce struggle for existence and which have their values in that struggle.

The bodily changes in great emotional excitement I have dealt with in detail elsewhere. They may be summarized here as illustrating another aspect of the natural defenses of the body than those we have thus far regarded.

In considering the homeostasis of blood sugar, oxygen supply, acid-base reactions and temperature, certain adaptive reactions were described which kept the body on an even course in spite of conditions which might have been deeply disturbing. It is remarkable that most of these reactions occur as the accompaniment of the powerful emotions of rage and fear. Respiration deepens, the heart beats more rapidly, the arterial pressure rises, the blood is shifted away from the stomach and intestines to the heart and central nervous system and the muscles, the processes in the alimentary canal cease, sugar is freed from the reserves in the liver, the spleen contracts and discharges its content of concentrated corpuscles, and adrenin is secreted from the adrenal medulla. The key to these marvelous transformations in the body is found in relating them to the natural accompaniments of fear and rage—running away in order to escape from danger, and attacking in order to be dominant. Whichever the action a life-or-death struggle may ensue.

The emotional responses just listed may reasonably be regarded as preparatory for struggle. They are adjustments which, so far as possible, put the organism in readiness for meeting the demands which will be made upon it. The secreted adrenin coöperates with sympathetic nerve impulses in calling forth stored glycogen from the liver, thus flooding the blood with sugar for the use of laboring muscles; it helps in distributing the blood in abundance to the heart, the brain, and the limbs (i.e., to the parts essential for intense physical effort) while taking it away from the inhibited organs in the abdomen; it quickly abolishes the effects of muscular fatigue so that the organism which can muster adrenin in the blood can restore to its tired muscles the same readiness to act which they had when fresh; and it renders the blood more rapidly coagulable. The increased respiration, the redistributed blood running at high pressure, and the more numerous red corpuscles set free from the spleen provide for essential oxygen and for riddance of acid waste, and make a setting for instantaneous and supreme action. In short, all these changes are directly serviceable in rendering the organism more effective in the violent display of energy which fear or rage may involve.

The remarkable arrangements which operate when we engage in hard muscular exercise, and which are prepared in an anticipatory manner in great emotional excitement, we can best understand by reference to racial history. For innumerable generations our ancestors had to meet the exigencies of existence by physical effort, perhaps in putting forth their utmost strength. The struggle for existence has been largely a nerve and muscle struggle. The organisms in which the adjustments were most rapid and most perfect had advantages over their opponents in which the adjustments were less so. The functional perfections had survival value, and we may reasonably regard the elaborate arrangements for mobilizing the bodily forces, which are displayed when intense muscular effort is required or anticipated, as the natural consequences of a natural selection.

Closely related to fear is pain. Indeed, fear has been defined as the premonition of pain. As a rule, pain is associated with the action of injurious agents, a fact well illustrated in cuts, burns and bruises. There are, to be sure, instances of very serious damage being done to the body—for example, in tuberculosis of the lungs—without any pain whatsoever; and there are instances, also, of severe pain, as in neuralgia, without corresponding danger to the integrity of the organism. These are exceptions, however, and the rule holds, that pain is a sign of harm and injury. By experience agents which injure and destroy, and which produce pain, become associated, so that our relations to them are conditioned by their effects. Thus pain saves us from repeating acts which in the end might make an end to life itself.

V

The foregoing review of some aspects of the natural defenses of the body must be looked upon as illustrative and suggestive rather than complete. Many more examples could be cited in each of the categories. The illustrations will serve, however, to indicate in what different ways the natural mechanisms of the body operate to shield it against harm and to repair and restore it in case of actual injury. Many of these safeguarding and healing processes are still mysterious. We are ignorant of the ways in which they are started and how they continue until the body is again intact. When we fully understand, we shall be able to use them.

It is noteworthy that in the resistance to attack by bacteria and their toxins, the protective agencies are present and active in the fluid matrix. Again we note that processes are set at work—just how, is not yet fully understood—which preserve the constancy of the blood. Antitoxins are produced which neutralize the toxins, and antibodies appear which play their part in destroying the invasive bacteria. And thus both the internal environment and the organism itself are preserved in their normal state. The changes which occur in emotional turmoil might, at first glance, be regarded as a gross disturbance of homeostasis. So they would be, by themselves; but they can be explained, I believe, only as preparatory for extreme muscular exertion. If that takes place, the changes in the fluid matrix at once become useful and are promptly counteracted by the effects of the exertion itself.

References

Cannon. Bodily Changes in Pain, Hunger, Fear and Rage. New York, 1929.
Cannon. The Mechanical Factors of Digestion. London, 1911, 57.
Cannon and Murphy. Annals of Surgery. 1906, xliii, 512.
Izquierdo and Cannon. Am. Journ. Physiol., 1928, lxxxiv, 545.
Zinsser. Resistance to Infectious Diseases, New York, 1931.

XV

THE MARGIN OF SAFETY IN BODILY STRUCTURE AND FUNCTIONS

I

IN 1907 Meltzer, in an important and suggestive paper, drew attention to a group of facts which he had gathered to throw light on the question whether our bodies are organized on a generous or on a narrowly limited plan. He pointed out that when an engineer estimates the weights which a bridge or beam must support, or the pressures to which a boiler will be subjected, he does not provide merely for those stresses in building the structure. The engineer multiplies his estimates by three, six or even by twenty, in order to make the structure thoroughly reliable. The greater strength of the material, above that calculated as necessary, measures what is known as a "factor of safety." How are our bodies built? was Meltzer's question. Are they set up with niggardly economy? Is barely enough provision made for keeping us intact? Or is there allowance for contingencies—have safety factors been introduced on which we may count in times of stress?

Already we have become acquainted with some evidence which answers these questions. We have seen that stores of carbohydrate, protein and fat are set aside in the body for use when supplies from the outside are not available. We have learned that although blood sugar is usually kept up to 90 or 100 milligrams per cent, it need not be that high. It can fall to 65 or 70, and sometimes lower, without producing disagreeable symptoms, and, as a rule, only the low level of 45-50 makes serious trouble. If 50 milligrams per cent is taken for the "threshold of adequate supply," as we may call it, or the "deficiency threshold," the margin of safety in blood sugar would be about 100 per cent.

Similar conditions prevail in the control of blood calcium. As we have noted, the normal concentration is about 10 milligrams per cent. Convulsions occur when the concentration is reduced about half; it may be reduced to 6 or 7, however, without danger of trouble. There is approximately the same margin for the calcium percentage as for the sugar percentage in the blood.

II

We have seen indications, too, that in the circulatory apparatus and its functions a large factor of safety is present. Although our normal systolic blood pressure is 110–120 millimeters of mercury, it may drop to 70–80 (i.e., about one-third) before reaching the critical level at which the volume flow to the tissues becomes insufficient. There is clearly a safety margin here. The quick restoration of approximately normal blood pressure, after a large percentage of the estimated blood volume, even up to 30 or 40 per cent, has been withdrawn, shows that the vasomotor apparatus is organized for security. As with other important arrangements in the organism, a series of devices assure the maintenance of an adequate blood flow. The vasomotor center is in the part of the brain (the medulla oblongata) which is nearest the spinal cord. When that is injured or destroyed, subsidiary centers soon assume control. And when they are eliminated, sympathetic ganglia take over the government. Finally, as Bradford Cannon has shown, all sympathetic influences can be excluded and then the vascular wall itself attends to the proper adjustment of the capacity of the vessels to the blood contained within them. Even in this, the last possible stage of reduction, therefore, blood pressure is held nearly up to the usual height.

We have had occasion to observe, also, that the heart is furnished with a large capability of meeting extra demands. Usually it beats at a moderate pace and puts forth a moderate volume of blood. But at any moment it is ready to contract twice as fast, put forth twice the amount of blood per beat, and against an arterial pressure which may be increased 30 or 40 per cent! It is a marvelously capable and adaptable organ, richly endowed with reserves of force.

III

In the respiratory functions, as well as in the circulatory, we find a large safety margin. Disease has proved that life may continue though a great part of the lungs has been destroyed. In some cases of pneumonia the lung on one side may become as solid as the liver without dangerously interfering with the oxygen supply to the body or with the elimination of carbon dioxide. This evidence is supported by the observation that collapse of the lung on one side so that it is no longer ventilated, or actual removal of half the pulmonary area, is endured without serious difficulty. In the lungs alone, therefore, the factor of safety is at least two.

In addition there is a much larger amount of oxygen transported from the lungs to the tissues than is ordinarily used. As we have previously noted, the blood which leaves the lungs with a load of approximately 18 volumes per cent of oxygen may return, when we are living quietly, still carrying 14 volumes per cent. In the blood, as it flowed past the masses of secluded cells, three and a half times as much oxygen was available as was actually taken for use.

The complex of adaptations which occur after hemorrhage or on ascent into the thin air of high altitudes, or when the oxygen-carrying power of the red corpuscles has been limited by carbon monoxide poisoning—the faster pulse, the increased blood pressure, the discharge of extra corpuscles from the spleen—must also be counted among the safety factors in the organization of the respiratory mechanism.

IV

It is a noteworthy fact in the construction of the body that many of its organs are paired. Are both of the paired organs needed for the continued existence and the efficiency of the organism? They are not. One of the two kidneys may be removed—indeed, two-thirds of each kidney may be taken—without serious disturbance of kidney function. The amount and composition of the urinary secretion is practically unchanged. This significant result is probably related to the important observation made by Richards of Philadelphia that at any one time many of the glomeruli of the kidney are not working—a condition which reveals directly a generous provision for special stress.

The safety margin is similar or even larger in other paired organs. The cortex of the adrenal glands is known to be necessary for life. If both glands are excised, death follows, usually within 36 hours. But if only one-tenth of the adrenal tissue is left the existence of the organism is not endangered. Again, the complete extirpation of the thyroid gland results in myxedema, with its lessened metabolism, slow reaction time, dry and thickened skin, and other abnormal consequences. Four-fifths of the thyroid substance, however, may be taken away without the appearance of any of these symptoms. The four small parathyroid glands, as already noted, are of the utmost importance in maintaining the proper calcium concentration in the blood. Their removal results in convulsions, coma and death, unless most skilful and intelligent nursing care is given. Elimination of at least two of the glands, however, is not followed by any perturbation at all.

In the nervous system the greatly elongated conducting cells are quite specially altered from their original, simple, roundish form. The more the cells are modified from this simple form the less capable they are of reproducing themselves. According to present evidence, if any of these long neurones are destroyed, there is no possibility of replacing them by division and growth of neighboring cells, as can be done in the liver, for example, when liver cells are locally killed. Thus, an injury to the nerve cells on the nasal side of the retina could never be repaired; a permanent blind spot would exist in the injured area. It would be compensated for, however, by the functioning of the corresponding lateral or temporal area in the other eye. Indeed, the factor of safety in the paired sense organs is at least two. The same relation holds true for the vagus nerves. Although section of both vagi causes pronounced digestive and respiratory difficulties, and is likely to lead to death in a few days from pneumonia, one nerve can be put out of action without causing any notable disturbance. Likewise severance of one of the great splanchnic nerves, which, like the vagi, distribute impulses over a wide range in the abdominal viscera, causes no observable impairment. In all respects the important functions of the abdominal organs which are concerned continue quite normally.

In view of the concept that nerve cells do not reproduce—that we have only one set of them, and if any are destroyed or injured they cannot be duplicated or repaired—we might suppose that the factor of safety in the brain would be nil. Dandy has reported, however, that when the growth of a tumor has required the operation, he has removed all of the right cerebral hemisphere above the basal ganglia with not only no danger to life but also with no appreciable change in mental characteristics or function. Likewise removal of both frontal lobes of the brain caused no notable effects. The patient was perfectly aware of time, place and person; the memory was unimpaired; he read, wrote, and passed mathematical tests accurately, and in conversation was not distinguishable from a normal individual. Nor was intelligence impaired by excision of the left occipital lobe or the lower half of the left temporal lobe. In fact, consciousness was permanently lost only when the area of the brain supplied by the left anterior cerebral artery was deprived of its blood supply. It is clear, therefore, that, so far as the functions of the brain that subserve conscious activity are concerned, a wide margin of safety has been provided. To be sure, removal of the cerebral hemisphere of one side does result in paralysis of the movements of the limbs of the opposite side of the body, but it does not effect muscles which contract on the two sides simultaneously—for example, the muscles of respiration and swallowing, which are essential for continued existence.

It is noteworthy that the brain and spinal cord, with their elemental duties of coördinating and controlling the activities of the organism, and with their peculiar lack of ability for structural regeneration if damaged, are specially protected by strong bony casings. The skull, though thin, is made of hard bone; and the spinal column, surrounding the spinal cord, though divided into vertebral segments which permit a certain amount of flexibility, is powerfully buttressed by ligaments and by surrounding muscles.

V

More striking, perhaps, than the safety factors in paired organs are those in the unpaired. The pancreas produces the internal secretion, insulin, which is required for the proper utilization of sugar by the organism. Complete removal of the pancreas, as previously noted, causes at once extreme diabetes. But four-fifths of the organ can be extirpated without ill effects; only one-fifth is necessary to furnish the insulin needed by the body.

Another instructive instance is offered by the liver. As Meltzer pointed out, it is an organ having many important functions. It plays a highly significant rôle in the metabolism of carbohydrates, fats and proteins. It protects the internal environment by supplying factors needed in the coagulation of the blood. It changes toxic ammonia compounds into relatively harmless urea. It excretes the pigments resulting from the breakdown of red blood corpuscles. It stands guard and prevents the entrance of metallic poisons from the alimentary canal into the general circulation and thus prevents their distribution throughout the organism. And it may have a central relation to the formation of red blood corpuscles. The liver is the busiest and most versatile organ in the body. Yet three-fourths of the liver can be taken out, and despite the variety and value of the functions which it performs, the loss does not induce symptoms which indicate any serious interference. The hepatic structure is obviously built greatly in excess of the normal requirements.

Again in the alimentary canal we find evidence that our bodily organs are not constructed on a pinched and skimpy scale. In operations for the treatment of disease or accident, most of the stomach has been removed, and yet digestion and nutrition have not been grossly impaired. About ten feet of small intestine have been taken out, and the patient has suffered no considerable ill effects. In many cases most of the large intestine has been cleared away, and the claim has been made that the result has been actually beneficial! It is evident that there is much more of the digestive tract than is needed for carrying on its functions.

One reason why the stomach can be removed without

greatly affecting the digestive process is that the pancreatic juice has a ferment which, like that of the gastric juice, is able to split the proteins of the food. This generous duplication of agencies is seen also in the arrangements for digesting starch; the salivary glands and the pancreas both produce starch-splitting ferments. Our six salivary glands, therefore, are not of primary importance for digestion; they can be wholly eliminated without interfering with the nutritive utilization of carbohydrates. And also fat, if finely emulsified (as it is in milk, for example), can be digested and absorbed, even in the absence of the fat-splitting ferment of the pancreatic juice, because there is a similar ferment in the gastric juice capable of attacking emulsified fat. Again we find, therefore, in the construction and workings of the alimentary canal, ample and liberal arrangements for assuring the existence of the organism.

VI

Many more instances might be cited to show that the various parts of our bodies are constructed with a wide margin of safety. As Meltzer wrote:

"The active tissues of most of the organs exceed greatly what is needed for the normal function of these organs. In some organs the surplus amounts to five, ten or even fifteen times the quantity representing the actual requirement. In the organs of reproduction the superabundance and waste of tissue for the sake of assuring the success of the function is marvelous. Furthermore, the potential energies with which some organs, like the heart, diaphragm, etc., are endowed are very abundant and exceed by far the needs of the activities of normal life. The mechanisms of many functions are doubled and trebled to insure the prompt working of the function. In many instances the function of one organ is assured by the ready assistance afforded by other organs. The continuance of the factors of safety is again protected by the mechanisms of self-repair peculiar to the living organism. We may, then, safely state that the structural provisions of the living organism are not built on the principle of economy. On the contrary, the superabundance of tissues and mechanisms indicates clearly that safety is the goal of the animal organism."

VII

The fathers of medicine made use of the expression, the "healing force of nature," the *vis medicatrix naturae.* It indicates, of course, recognition of the fact that processes of repair after injury, and of restoration to health after disease, go on quite independent of any treatment which a physician may give. All that I have done thus far in reviewing the various protective and stabilizing devices of the body is to present a modern interpretation of the natural *vis medicatrix.* As we have seen, there are various ways in which through many years the normal state of the organism is maintained or its disturbed balance reëstablished by automatic physiological reactions. The numerous methods which throughout human history have been employed to cure disease, from beating a tom-tom to the royal touch and the use of prayer, have all been justified by the fact that persons who were ill became well under the treatment. Only in recent times have any considerable number of persons been willing to test the efficacy of the natural processes alone and to observe that they are potent factors, working for health. If the body can largely care for itself, however, what is the use of a physician?

In the first place, the well-trained physician is acquainted with the possibilities and limitations of self-regulation and self-repair in the body. He is instructed in that knowledge and employs it not only for his own intelligent action but also as a means of encouragement for the patient who looks to him for counsel. For example, external heat, plus that produced by the working parts, may be so great as to run the body temperature up to a dangerous height—i.e., the adaptive mechanisms may be overwhelmed unless such external aid as the alert physician can give is immediately rendered. Or, to take another example, great fear, with its attendant internal preparations for struggle, may be serviceable in wild life when the need for physical effort is imminent, but in the circumstances of civilized existence it may be the occasion for baneful disturbance of vitally important functions. These are facts which the informed physician understands and can explain in ways which are helpful and curative.

Again, the physician realizes better than the layman that many of the remarkable capacities of the organism for self-adjustment require *time*—all of the processes of repair belong in that class—and that they can play an important rôle in restoring the organism to efficiency only if they are given the chance which time provides. The wise physician, therefore, insists on conditions which permit only such activities as are necessary until lost or injured parts have been rebuilt, strengthened, or compensated for.

Furthermore, the physician realizes that he has at his command therapeutic agents with which he can support or replace the physiological self-righting or self-protective processes we have been considering. When he gives insulin for diabetes, for instance, he knows that his treatment is serving in a natural manner to perform a natural function which has broken down, and that neither bone setting nor mental therapy can be useful in its place. Or when he gives thyroxin for myxedema or cretinism, again he is aware that he is using a physiological factor for a physiological defect. And antitoxin, he well understands, is a means of helping the defensive reactions which are regularly a part of the body's self-protection. He appreciates the fact, also, that a mode of treatment used on a sick person to restore the normal state is commonly more potent than when used on a well person. Thus, cold applications will reduce a high fever, whereas the same applications will not reduce a normal temperature. Or a dose of thyroxin which will markedly raise a metabolism which is low because of thyroid deficiency will have little or no effect in raising the normal metabolism. Or a given dose of insulin has much greater influence on a severe case of diabetes than it has on a mild case. The physician, then, plays his part in making effective the self-regulating adjustments of the body that have been disordered or that are in need of reinforcement, understanding that, as a rule, nature herself is working with the curative agencies which he applies.

Finally, a great service which the physician renders is that of bringing hope and good cheer to his patients. That alone justifies his presence. He has seen at work in many cases the restorative processes of the organism. In the facts which we have surveyed we have become acquainted with good reasons for extending hope and cheer to the sick, reasons based on the ample evidence that in the body there are admirable devices for maintaining its stability against disturbing internal and external conditions, marvelous provisions for protecting its integrity against foes, both wild beasts and microscopic germs, and very liberal margins of structural strength and functional capacity beyond the ordinary requirements. When we are afflicted and our bodily resources seem low, we should think of these powers of protection and healing which are ready to work for the bodily welfare.

References

Cannon, B. Am. Journ. Physiol., 1931, xcvii, 592.
Dandy. Am. Journ. Physiol., 1930, xciii, 643.
Meltzer. "The Harvey Lectures," 1906–07, i, 139.
Richards. "The Harvey Lectures," 1920–21, xvi, 163.

XVI

THE GENERAL FUNCTIONS OF THE TWO GRAND DIVISIONS OF THE NERVOUS SYSTEM

I

WE HAVE now completed consideration of such aspects of homeostasis as are, in present knowledge, fairly well defined, and we have become acquainted with some of the natural defenses of the body, as well as with safety factors in its structure and functions. The possibility of obtaining further insight into the organization which makes for resilience and endurance in spite of the fell blows of circumstance lies in an examination of the ways in which stability is achieved.

As a basis for the attempt to understand the management of steady states in the fluid matrix, we must have in mind the broad outlines of the functions of the nervous system. According to its functions the nervous system may be separated into two great divisions: that acting outwardly, in relation to the external environment of the individual; and that acting inwardly, on the viscera, and governing principally the internal environment. Our interest will be directed especially to the latter division, but in normal existence the two divisions are not separable.

II

We are all familiar with the gross appearance of the brain and spinal cord and their situation in the skull and back bone respectively. We know also that nerve fibers connect the brain and spinal cord with every point on the surface of the body and with every muscle which we move. The fibers leading in from the surface, the so-called "sensory fibers," are stimulated through the mediation of sense organs, or "receptors," or "exteroceptors" (to distinguish them from internal sense organs). These exteroceptors are as a rule cells, or groups or layers of cells, which are peculiarly sensitive to different types of external agents. Among these sensitive agents are touch corpuscles, affected by contacts; temperature endings, influenced by heat and cold; the olfactory area in the nose, excited by chemical substances in the air; the taste buds on the tongue, stimulated by certain chemical agents when dissolved in water; the internal ear, for the reception of aerial vibrations transmitted to it through a bony lever; and the eye, exquisitely impressionable by the waves of light. Some of these receptors require immediate presence of the stimulating agent, as, for example, those for touch and taste. Others, e.g., those for smell, hearing and sight, are responsive to changes at more or less remote distances from the body, and they are known, therefore, as "distance receptors." By means of these sensitive organs on the surface of the body we are able to become acquainted with all sorts and conditions of objects in our surroundings, from the texture of a piece of cloth to the characteristics of stars hundreds of thousands of light years away.

From each of these exteroceptors the nerve fibers which lead inward carry to the brain or spinal cord—which constitute the "central nervous system"—nerve impulses which subserve sensation. They announce the moment when an external agent stimulates a receptor, and the degree and character of the stimulation.

From the central nervous system other nerve fibers lead outward to the muscles which move the levers of the skeleton, and also to other muscles, such as those of the face, which are not attached to the bones. The muscles innervated by these "motor nerves" are known as "effectors" or "effector organs." The "motor nerves," of course, are useless unless they can operate the muscles. And, as illustrated by the paralytic, the muscles likewise are useless unless they are made to act by nerve impulses. The outwardly acting division of the nervous system, therefore, must be regarded as naturally a neuro-muscular organization.

The brain and spinal cord are constituted of an enormously intricate arrangement of nerve tracts, which can connect every station in an afferent path from an exteroceptor with any station in the efferent path to an effector. Thus the reaction or the response of the individual to a given stimulus (e.g., an itching sensation from the left shoulder) or to a group of stimuli (e.g., a runaway horse) which present a complex external situation, can be properly adjusted, so that the behavior that results is adapted to the requirements.

III

The simplest response is the reflex, an immediate discharge of motor impulses, in answer to stimulation of a receptor, with consequent muscular contraction. A sneeze, a cough, the quick winking of the eyes, and the maintenance of posture, are illustrations of reflexes. Usually they are protective in character. They are quite involuntary and are not associated with any mental activity. Somewhat more complicated are the reactions related to feelings and emotions, such as laughing, weeping, and the typical attitudes of anger and fear. Like reflexes these reactions are not learned; they can be observed in the early days of life, too soon to be imitated. The reflexes are managed mainly in the spinal cord and in the lower parts of the brain (i.e., the medulla oblongata and the mesencephalon, see fig. 33) which are most nearly connected with the cord. The reactions of an instinctive type, which resemble reflexes, have their central organization in the basal part of the brain. The physiological manifestations of anger, for example, will develop quite fully after all the parts above the diencephalon have been removed, as Bard has demonstrated. The phenomena promptly vanish on additional removal of the diencephalon.

The organ for associative memory, and for all the com-

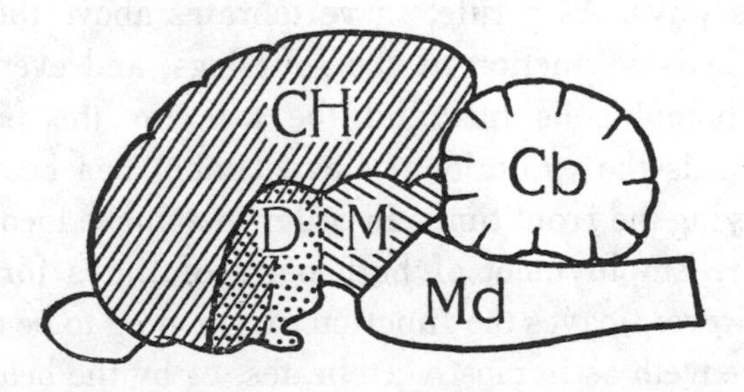

Fig. 33. Median section of the mammalian brain. CH, cerebral hemispheres; D, diencephalon (indicated by dots); M, mesencephalon (mid-brain); Md. medulla oblongata: Cb. cerebellum.

plicated adjustments of muscular reaction to our surroundings is the cerebral cortex, the outer portion of the cerebral hemispheres (see fig. 33). In other parts of our central nervous system we may not be very unlike lower animals, but in the cerebral hemispheres we are almost incomparably different. In this region the connections with receptors and with muscular effectors are more numerous and detailed in man than in any lower animal, and the interrelations between the central stations for the receptors and effectors are vastly more intricate. On this elaborate organization of the nerve connections in the cerebral cortex is based the much superior intelligence of man as compared with other mammals, and his ability to modify the external environment in extraordinary ways.

IV

The fundamental problems of continuing existence and of propagating the race, that must be met by organized beings, require two primary activities, locomotion and grasping. These activities have been carried on in various ingenious ways. As a rule, in vertebrates above the fishes locomotion is a function of the four legs, and even fishes use the homologous fins and the tail for the purpose. Among birds the evolutionary experiment has been tried of modifying the front limbs in order to achieve locomotion in the air. Employment of both pairs of limbs for movement, however, leaves the function of grasping to be carried out by the teeth as in most vertebrates, or by the beak as in birds, or by the nose as in the elephant. There are, of course, indications of special or double use of parts; the frog clasps with his short legs, the bear hugs as he stands, the squirrel holds nuts with his fore paws, and monkeys and apes have hand-like feet which they use not only for progression but also prehension. In man, however, fairly complete division of function is reached in an assignment of locomotion to the hind limbs and prehension to the fore limbs and hands. The development of the remarkable ability of the hands to perform all manner of actions is apparently associated with the development of the complex organization of the fore brain and the cerebral cortex. By means of hands, tools and instruments have been made—picks, saws, brushes, scalpels, lathes, steam hammers, and what not else!—which prodigiously amplify both the strength and the delicacy of manual operations. Tools and instruments, in turn, have been used to contrive devices which immensely extend the range of our receptors—the microscope for seeing little things, the telescope for seeing things far away, and the radio receiver and amplifier for hearing from remote sources the minute electrical pulses in the air.

The cerebrospinal nervous system, elaborately outfitted with sensitive exteroceptors and with a multitude of muscles which operate to any desired degree and in almost any direction numerous bony levers, is arranged for altering the external environment or the position of the organism in that environment by laboring, running or fighting. These outwardly directed activities may properly be designated as *exterofective*, and the so-called "voluntary nervous system" may be quite exactly designated as the exterofective division of the nervous system. We have already learned, however, that exterofective activities must produce coincident changes in the internal environment, as in the utilization of blood sugar and the discharge of acid waste and extra heat into the streaming blood. In these circumstances the "involuntary nervous system" plays its part by influencing the heart, and the muscles and glands of other viscera, in such ways as will preserve the fitness of the internal environment for continued exterofective action. This inwardly directed functioning of the involuntary nervous system justifies calling it the *interofective* system. We shall now consider its organization.

V

The interofective division of the nervous system, besides being called the "involuntary" system, is known also as the "vegetative" or the "autonomic" system: "vegetative" because it is concerned largely with the nutrition of the organism rather than with the animal functions of locomotion and prehension; and "autonomic" because it acts automatically, without direction from the cerebral cortex. The autonomic system has the task of adjusting the functions of the viscera to the advantage of the organism as a whole. It has a number of characteristics which we need to know in order to see the significance of certain experiments which will be described later.

First, the parts of the viscera innervated by autonomic nerve fibers are smooth muscles ("smooth," to distinguish them from the striated muscles attached to the skeleton) and glands. The smooth muscle cells are found at the root of the hairs, in the coats of blood vessels, encircling the bronchioles of the lungs, and in the walls of hollow structures like the stomach and intestines, the bladder and the

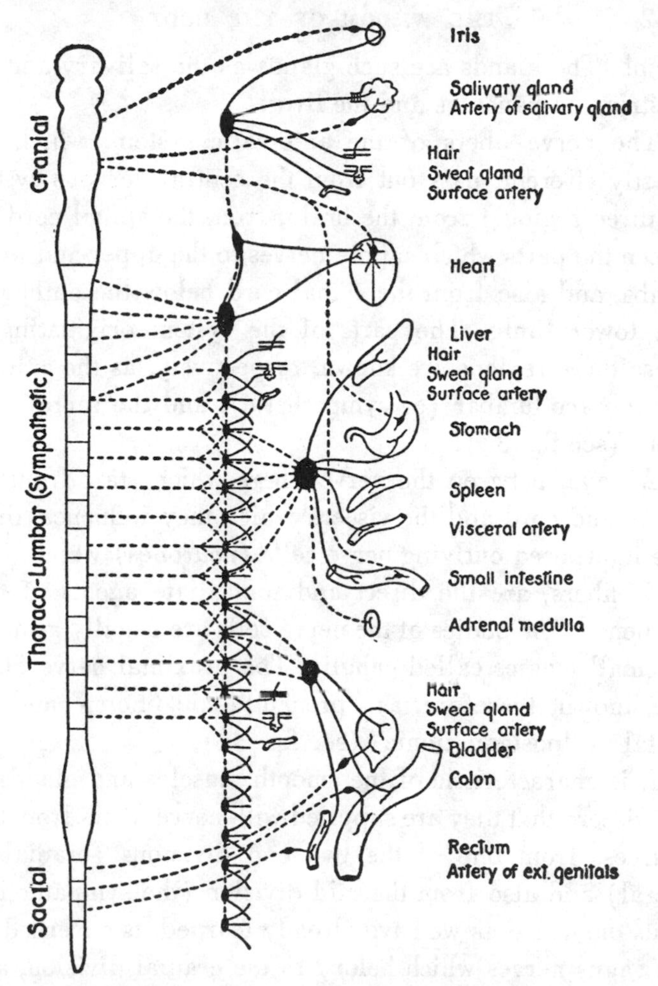

Fig. 34. Diagram of the general arrangement of the autonomic nervous system. The brain and spinal cord are represented at the left. The nerves of the somatic system are not shown. The preganglionic fibers are in broken lines, the postganglionic in solid lines. Further description in text. (From Bard [after Cannon], "Foundations of Experimental Psychology." Courtesy of the Clark University Press.)

womb. The glands are such glands as the salivary and the gastric, the pancreas and the liver.

The nerve fibers of the automatic system, which are mostly efferent, pass out from the central nervous system in three regions: from the brain, from the spinal cord between the parts which supply nerves to the upper and lower limbs, and also from the spinal cord below the outflow to the lower limbs. The parts of the system originating in these three regions are known, respectively, as the cranial, the thoraco-lumbar (or sympathetic), and the sacral divisions (see fig. 34).

Always, between the nerve fibers which start from the brain and cord and the viscera which they influence, there are interposed outlying nerve cells (neurones) which, with their fibers, are the direct and immediate agents of that influence. The bodies of the nerve cells are usually grouped in small masses called ganglia. The proximal nerve fibers are known, therefore, as "preganglionic fibers," and the distal as "postganglionic" (see fig. 34).

It is characteristic of the smooth muscles and glands of the viscera that they are supplied with nerve fibers from two sources, from one of the two end divisions (cranial or sacral) and also from the mid division (the sympathetic). Thus the heart, as we have already learned, is governed by the vagus nerves which belong to the cranial division, and also by the accelerator fibers which come from the sympathetic division.

The representatives of the two divisions which supply any viscus are, as a rule, opposed in their effects. The vagus acts to check the heart rate, the sympathetic to speed it up; the vagus serves to increase the moderate contraction or tone of the gastric muscle, the sympathetic to decrease it. Numerous other similar illustrations might be added.

The cranial and the sacral divisions are peculiar in that their preganglionic fibers pass into or reach close to the viscera which they affect before they meet the relay afforded by the outlying neurones. In other words, the ganglia and the postganglionic fibers of these neurones lie in or near the effector organ (see fig. 34). Conditions are quite different in the sympathetic division.

VI

Except in the neighborhood of the large branches of the aorta in the abdomen (where there are special ganglia for the abdominal viscera), the ganglia of the outlying nerve cells of the sympathetic division lie in two chains on either side of the vertebral column, from the superior cervical ganglion high in the neck to coalesced ganglia in the pelvis. In the chest and abdomen these ganglia are found at the back of each cavity close to the bodies of the vertebrae. They are bound together by preganglionic fibers which come out from the spinal cord in a regular series between the vertebrae, pass to the nearest ganglion, connect with the outlying neurones of that ganglion, and then reach downward or upward to other ganglia in the chain, connecting with the neurones in each. This arrangement provides a highly effective device for multiplying contacts and thereby increasing the channels of distribution. Ranson and Billingsley have counted in the cat the number of preganglionic fibers in one cervical sympathetic trunk (composed of preganglionic fibers) and the number of cell bodies in its superior cervical ganglion, and they report that the ratio is as 1 to 32. Obviously, if a similar relation exists generally, any single preganglionic fiber would bear impulses having extensive effects because of the multiplicity of its connections with the final distributing agents. And the bridging of several ganglia by preganglionic fibers in the thoracic and abdominal regions would give the bridging fibers control over the areas supplied from all the ganglia.

The bridging or connecting fibers, just referred to, by reaching some distance up or down the chain of ganglia on each side of the spinal column, overlap one another. This results in a diffuse, widespread discharge of nerve impulses through sympathetic channels, as contrasted with the limited, sharply directed discharge to specific organs in the functioning of the cranial and sacral divisions.

The view that the sympathetic division is organized for diffuse effects is supported by common observation. When a cat, for example, is subjected to excitement or to exposure to cold—conditions which call the sympathetic into action—its hairs are erected from head to tail tip. Deeper observation reveals that the general erection of the hair is indicative of other general effects on large systems, such as inhibition of activities throughout the digestive canal and constriction of blood vessels in all regions of the body. On the other hand, the different functions of the *cranial* division are not bound to be performed simultaneously but may be performed separately. Thus the impulses which run along cranial autonomic pathways and cause constriction of the iris when the light is too strong for the eyes, are not necessarily associated with impulses which cause increased flow of saliva or slowing of the heart or increased tone of the gastro-intestinal tract. These effects of the cranial division are distinct, just as the preganglionic fibers to the several organs are distinct.

The view that the sympathetic division is organized for diffuse discharge of nerve impulses is confirmed also by the fact that sympathetic impulses evoke a secretion of adrenin from the medullary portion of the adrenal glands. This substance, when injected into the blood stream, produces, as a rule, in organs innervated by sympathetic nerves, the same effects as are produced by the nerve impulses. Since secreted adrenin has a general distribution in the blood stream, the sympathetic division, even if it did not have diffuse effects because of the way its fibers are arranged, could have such effects by the action of adrenin. The two factors, however—the nerve impulses and the chemical agent in the circulation which simulates their effects—commonly work together, as a sympathico-adrenal system, to produce widespread changes in smooth muscles and glands throughout the organism.

The question has been raised as to whether the extra adrenin discharged into the circulation on special occasions is, in fact, present in the blood in sufficient amount to influence the viscera. There are many observations which permit this question to be answered affirmatively. We have already had occasion to note the effect on the denervated heart of adrenin which was secreted in a natural manner by the adrenal glands when they were stimulated by hypoglycemia and cold (see figs. 18 and 27). Similar proof has been obtained on the denervated salivary gland, iris, and kidney, and also by means of a crossed circulation from the adrenal vein of one animal to the general circulation of another. By all these experimental methods proof has been obtained that adrenin set free in the blood is distributed in a concentration which influences effectively the distant organs which have been used for purposes of test.

Although adrenin is distributed by the blood and can therefore coöperate with sympathetic impulses to produce the same effects which they produce, what is the evidence that its coöperation is useful? First, there are indications that circulating adrenin prolongs the effects of sympathetic activity. Britton and I have reported a continued acceleration of the denervated heart, during nearly a half hour after one minute of excitement, although the animal, after the momentary disturbance, was resting serenely on a cushion (see fig. 35). When conditions require prolonged activity, therefore, the secreted adrenin would be advantageous. Furthermore, there are observations that in some respects the secreted adrenin has an efficacy far beyond that of the sympathetic nerve impulses alone. For example, asphyxia, reflex stimulation, normal emotional excitement, and the sham rage which follows removal of the cortex from the cerebral hemispheres, are all associated with hyperglycemia. This is fully as great in animals with the liver nerves severed and with the adrenal glands undisturbed as it is in animals normal in that respect. On the other hand, if the adrenal glands are removed or inactivated and the liver nerves are left intact, the hyperglycemia, under the same experimental conditions, is diminished or entirely lacking (see p. 117). Quite pos-

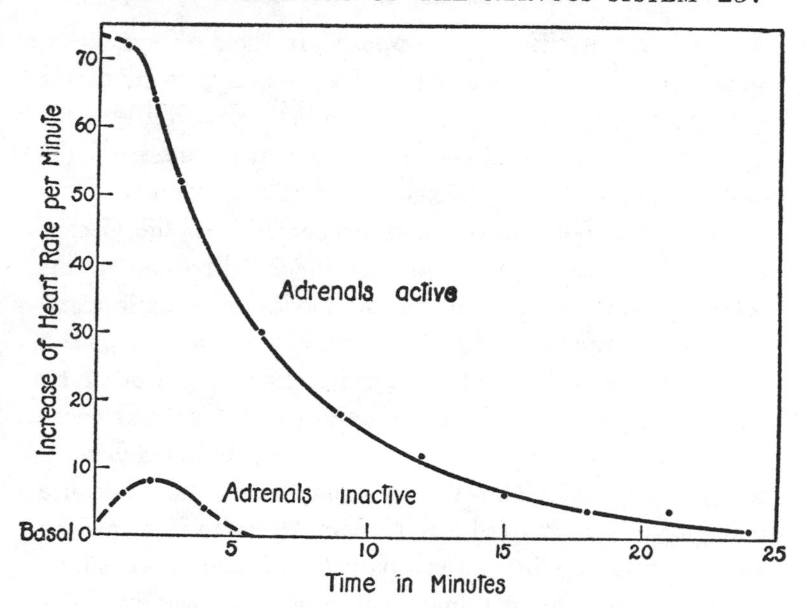

Fig. 35. Long persistence of the faster heart rate, when the adrenals were active, after the animal (cat) was excited by a barking dog for one minute. Only slight increase of heart rate and its relatively rapid subsidence, after adrenal inactivation, though the cat was excited twice as long as before. In both instances the animal was promptly removed from the cage after the excitement, and during the period of recovery rested quietly on a cushion.

sibly the action of adrenin in causing a faster coagulation of the blood and a faster metabolism, as well as a peculiar influence which it has in abolishing the effects of muscular fatigue, is quite independent of any coöperation with sympathetic nerve impulses.

VII

Associated with the double nerve supply to the viscera and with the interesting fact that, as a rule, the two nerves meeting in any viscus are opposed in their effects, is an important possibility of adjusted action. By means of the general *diffuse* effect of the sympathetic division and the opposite *particular* effect of the several nerves of the cranial and sacral divisions, every variety of change is provided for, both locally and temporally. *All* the viscera can be influenced *simultaneously* in one direction or the other by varying, up or down, the moderate or tonic activity of the sympathetic division. And any *special* viscus can be separately influenced in one direction or the other by varying, up or down, the tonic activity of the special nerve of the opposed cranial or sacral division that reaches directly to the viscus. Thus the heart may beat rapidly because that effect is part of the total complex of effects on the viscera produced by the sympathetic in emotional excitement, for example, or it may beat rapidly without extensive involvement of other viscera because of a lessening of vagal inhibition, as at the start of a muscular movement. The sympathetic is like the loud and soft pedals, modulating all the notes together; the cranial and sacral innervations are like the separate keys. When we consider that in emergencies the sympathetic functions in a great variety of ways to serve the organism as a whole, the importance of its arrangement for simultaneous and unified action becomes evident.

VIII

The use of the terms "involuntary" and "autonomic" to characterize the interofective division of the nervous system implies a dissociation of that system from such control as we exercise over skeletal muscles. We cannot, by act of will, inhibit the movements of the stomach and intestines, slow the beating of the heart or liberate sugar from the liver. Yet, at appropriate times all these changes, and many more, are effected by action of autonomic nerve impulses. The reactions in the autonomic system are evidently in close relation to simple reflexes, being prompt, inborn, unwilled, and purposive. We are likely, therefore, to obtain insight into the functions of the divisions of the autonomic system by considering the special circumstances under which each one of them is brought into action and also the special effects which each produces. The following summary accords with suggestions which I put forward in 1914.

The function of the *sacral* division is in the main to empty hollow organs which are periodically filled. Thus, sacral autonomic impulses cause contraction of the rectum and distal colon and also contraction of the urinary bladder and possibly other receiving structures. In the best understood instances the effects are induced reflexly by a stretching of the tonically contracted viscera by their accumulating contents. Although this reflex response to distension is probably the commonest mode of action of the sacral division there are other ways in which it is stimulated to activity. Great emotion—fear, for example—which is attended by nervous discharges through the sympathetic division, may be attended also by discharges through sacral nerve channels. The involuntary voiding of the bladder and lower large intestine in times of intense excitement is a well-known phenomenon in man and in other animals. And special affective states can markedly influcnce the functions of the organs of reproduction. Because in some cases the hollow organs of the pelvic region must discharge through sphincters composed of striated muscle, there is, to a certain degree, cortical or voluntary control of their emptying. Also by voluntary contraction of the abdominal muscles the process of emptying can be hastened. But in their essential features the functions of the sacral division are automatic. In fact, as shown by wounded men in the Great War, these functions can be performed after that part of the spinal cord in which the centers are located is quite isolated from the brain.

The functions of the *cranial* division, like those of the sacral, can be summarized in the statement that they are mainly a group of reflexes, protective, conservative and up-building in their service. By narrowing the pupil of the eye the cranial division protects the retina from excessive light. By providing for the flow of saliva and gastric juice and by establishing in the muscles of the wall of the alimentary canal the tonic state which is necessary for their periodic contractions, it assures the fundamental basis for proper digestion and absorption of the energy-yielding material required for all bodily activity. By vagus control of insulin secretion it may play a rôle in the storage of glycogen in the liver. Further evidence of the conservative value of the cranial autonomic influence is found in the provision of time for rest and recuperation of cardiac muscle by tonic vagal slowing of the heart rate.

The cranial division, like the sacral, not only has uncomplicated reflexes, such as the narrowing of the pupil, but also it has activities associated with affective states. The pleasurable taste and smell of food are accompanied by the so-called "psychic" secretion of the digestive juices and by a tonic contraction of the stomach and intestines. Furthermore, the two divisions—sacral and cranial—are similar in being largely subject to interference by the movement of striated muscle. Just as contraction of the bladder and rectum can be aided or checked by nerve impulses from the cerebral cortex, the reactions of the pupil to light or to distance can be induced or modified by voluntary acts. Indeed, as a rule, the workings of the sacral and cranial divisions involve the coöperation of the cerebrospinal nervous system to a much greater degree than do the workings of the sympathetic division, because they are much concerned with external orifices surrounded by striated muscle.

IX

As an interofective system exerting its influence on the activities of the viscera the autonomic must necessarily be intimately involved in the preservation of that stability and constancy of the internal economy of the organism which we have called homeostasis. In fact, many of the functions of the three divisions can be reasonably interpreted as directed towards homeostatic conditions. The sacral division, as a group of reflexes for emptying hollow organs which periodically fill, preserves constancy by getting rid of waste and by discharging accumulations which disturb or limit the free action of the individual. Incidental thereto the race is assured continuance. The cranial division, as a group of reflexes which are protective and conservative in their effects, is more obviously concerned with the preservation of uniformity in the organism. Insofar as its main services are involved—those of regulating digestion and absorption of food—it is of primary importance for the maintenance of steady states.

The sacral and cranial divisions of the interofective system, however, operate only indirectly and somewhat remotely to assure a constant state. It is the middle or thoraco-lumbar division which acts promptly and directly to prevent serious changes of the internal environment. So important for homeostasis are the uses of this division that it deserves special and detailed consideration.

References

Bard. Am. Journ. Physiol., 1928, lxxxiv, 490.
Cannon. Am. Journ. Psychol., 1914, xxv, 256.
Cannon. The Autonomic Nervous System, an Interpretation. The Lancet, 1930, i, 1109.
Cannon and Britton. Am. Journ. Physiol., 1927, lxxix, 433.
Ranson and Billingsley. Journ. Comp. Neurol., 1918, xxix, 305.

XVII

THE RÔLE OF THE SYMPATHICO-ADRENAL SYSTEM IN HOMEOSTASIS

I

IN AN earlier chapter I followed Claude Bernard in emphasizing the fact that we do not live in the atmosphere which surrounds us. We are separated from that atmosphere by a layer of dead cells or by a film of mucus or of salt solution. All that is alive within these lifeless surfaces is immersed in the fluids of the body, the blood and the lymph, which form an internal environment. Striking instances of the dangers which arise when the internal environment, or fluid matrix, is markedly altered, clearly demonstrate the primary importance of keeping it as constant as possible. With clear vision Claude Bernard saw that the preservation of constancy is the condition for free and independent life. As we go about we carry with us our internal climate. Consequently, changes in temperature, for example, or in moisture or in the oxygen percentage, in the world around us, unless they are extreme, have little effect on the realm wherein we live. But this comfortable constancy, or homeostasis, is not secured without the operation of nice devices which normally are always ready to prevent disaster. Both without and within the organism conditions may arise which, if permitted to prevail, would quickly produce profound disturbances.

For example, external injury may open blood vessels and allow the fluid matrix itself to escape. Coagulation of the blood appears, however, as a conservative agency. And by action of the sympathico-adrenal system the efficacy of coagulation in stopping hemorrhage is increased by acceleration of the clotting process. If perchance hemorrhage occurs to such a degree that there is considerable loss of blood, again the sympathico-adrenal system becomes active and constricts the peripheral vessels, thus diminishing the flow in regions where bleeding is likely to be prominent and also assuring a continuous supply of blood to essential and constantly active organs, the heart and the brain.

Again, there may be external cold which withdraws heat from the organism and thereby threatens a lowering of the temperature of the internal environment. The sympathico-adrenal system promptly acts to avert that danger. It constricts peripheral blood vessels and lessens the exposure of warm blood to the surface. It causes erection of hairs and feathers, in animals provided with these means of protection, so that a layer of poorly conducting air surrounds the body. It liberates into the circulation adrenin which increases the speed of oxidative processes in the body at a time when extra heat is needed to keep the temperature from falling.

Or there may be a reduced oxygen supply to the organism, either because of resort to high altitudes, or because of the action of a poison, such as carbon monoxide. Once more the sympathico-adrenal system comes to the support of homeostasis. The heart is made to pump more rapidly, the vascular area in the splanchnic region is constricted, and in consequence the blood flow is faster in essential structures. Furthermore, the spleen is made to contract and discharge myriads of additional red corpuscles into the service of the organism. By these functions of the sympathico-adrenal system what might be a serious disturbance is minimized to an important degree.

External conditions, however, are not the only factors which affect the internal environment. The activity of the body itself may upset homeostasis; and, if not guarded against, profound disorders may result. As we have learned, among the disturbing internal conditions vigorous muscular work stands forth prominently. Let us review the changes caused by extreme physical effort and see how they are prevented from altering the fluid matrix of the organism.

II

The activity of muscles is associated with the using up of sugar. Recall that prolonged running markedly reduces blood sugar, and may deplete the liver of its store of glycogen. Recall, also, that reduction of blood sugar to about 45 milligrams per cent, either by extirpation of the liver or by use of insulin, results in convulsions; and should the reduction continue, coma and death would follow. Under normal conditions, however, before the convulsive stage is reached the sympathico-adrenal system is brought into operation and sugar is liberated from the liver. And so much sugar is thus liberated that, if the agency producing hypoglycemia is not overwhelmingly effective, the danger is passed and the sympathico-adrenal system then subsides into inactivity. Thus, in spite of great demands for sugar from laboring muscles, the glycemic level is maintained above the region of serious danger.

Again, vigorous muscular activity is associated with the development of heat. As we have noted, the amount of heat produced when the great muscle masses of the body are employed in prolonged and supreme effort may be almost incredibly large. Unless this extra heat is dissipated, the temperature of the internal environment may rise to a perilous degree. Temperatures as high as 40.5° C. (104.9° F.) in man have been reported as consequent on exertion. Two degrees more, if prolonged, may bring disaster because of effects on sensitive nerve cells in the brain. For protection against this danger the sympathetic system is called into service. Peripheral vessels are relaxed, sweating becomes profuse, and the discharge of heat is greatly accelerated. Thus a rise of temperature is quickly met and counteracted, and the harmful effects of high temperature are obviated.

Furthermore, when one engages in strenuous muscular work there is a tendency for the state of the blood to shift towards the acid side of its nearly neutral reaction, an effect due, as we have learned, to increase of carbonic and of lactic acid, resulting from neuromuscular activity. The production of acid is not without danger. As is well known, in pathological states a condition of "acidosis" may develop to such a degree that the functions of the nervous system are impaired, and, if the situation is not rectified, coma and death may supervene. Long before that state is reached, as a result of muscular effort, corrective and protective processes are started and vigorously maintained. The important features in the strategy are, first, an absorption of extra acid in the buffer substances of the blood; second, a prompt supply of extra oxygen to burn the non-volatile lactic acid (which otherwise persists and makes difficulty) to volatile carbonic acid, which can be rapidly discharged; and third, an acceleration of breathing so that carbon dioxide is driven away from and extra oxygen is drawn into the lungs. In short, the circulatory and respiratory mechanisms work at their maximal capacity. Once more the sympathico-adrenal system steps in to save the fluid matrix from grave disturbance. The circulatory adjustments—constriction of the splanchnic vessels, acceleration of the heart, discharge of extra corpuscles from the spleen—are all made by means of the sympathico-adrenal system. And in addition, this system probably plays a rôle in facilitating the respiratory processes, for it can quickly and effectively cause dilation of the bronchioles and thus reduce the frictional resistance to the to-and-fro movement of the respired air. By these various services the sympathico-adrenal system plays a most important rôle in preserving the homeostasis of the acid-base relationship in the blood, at times when it is dangerously threatened.

The foregoing survey has shown that states of the *external* environment, and also responses of the organism itself to situations in the external environment, are associated with disturbances of the *internal* environment. This personal, individual climate, which we carry with us, must not greatly change if we are to continue to be effective. For constancy of the internal environment, therefore, every change in the outer world, and every considerable move in relation to the outer world, must be attended by a rectifying process in the inner world of the organism. The chief agency in this rectifying process, as we have noted in many illustrations, is the sympathetic division of the autonomic system.

III

Having now laid emphasis on the great importance of the sympathetic, or sympathico-adrenal mechanism, I propose next to describe the physiology of animals from which Newton, Moore and I removed this part of the autonomic nervous system. We have already learned that one of the characteristic features of the true sympathetic neurones is the disposition of their cell bodies in a chain of ganglia lying along either side of the spinal column, from the superior cervical ganglion high in the neck to the fused ganglia in the pelvis (see fig. 34). The ganglia are connected to the spinal cord from the first thoracic to the second or third lumbar segments of the spinal cord. The aggregation of the nutritive centers (the cell bodies) of the outlying neurones in ganglionic chains, especially in the chest and abdomen, where direct connections with the spinal cord are present, permits a thorough extirpation of the system. If the ganglia are removed, the postganglionic fibers must degenerate. And, as Langley has shown, the preganglionic fibers, coming from the central nervous system, are incapable of making effective connection with the organs normally innervated by postganglionic fibers. Removal of the sympathetic chains, therefore, must necessarily result in a permanent disconnection of these organs from central nervous control via sympathetic channels.

In removing the sympathetic chains we used two methods. First, we took out the cervical, the thoracic, and the abdominal portions separately. One of many animals that survived in good health such an operation is shown in figure 38. It is a photograph of a cat after removal of the semilunar ganglia, and the last portion of the sympathetic chain. The portions which were removed were mounted on a card and the card was photographed at the same time with the cat which formerly possessed them. Below each part is the date of its removal. The defect of this method of operation lay in the occasional failure to excise one or two ganglia in the lower thorax, close to the diaphragm. To avoid that possibility we removed the thoracic and abdominal chains, from the stellate ganglia to the ganglia in the pelvis, in an unbroken condition. In figure 36 are illustrations of

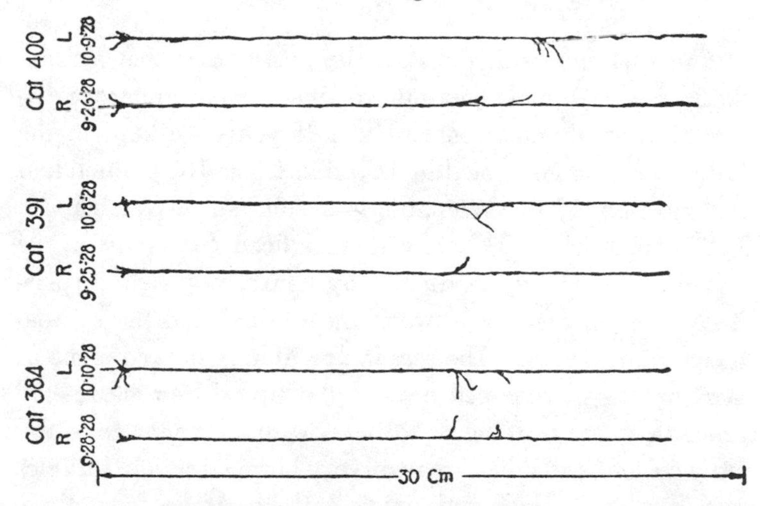

Fig. 36. Examples of thoracic and abdominal sympathetic chains removed intact. They were dried and with collodion were mounted on the card before being photographed.

such complete removal. Since the preganglionic fibers for the cervical sympathetic ganglia pass through the stellate ganglion on either side, the removal of the stellates necessarily disconnects the cervical ganglia from the central nervous system. That the cervical ganglia do not function in these circumstances is shown in figure 38. The *right* thoracic and abdominal sympathetic chain had been taken out entirely. When the animal was cooled the innervated hairs on the *left* side stood erect. Note that the hairs of the *head* on the right side, though still innervated from the isolated cervical ganglia, did not respond to cold. It is clear, therefore, that removal of the thoracic and abdominal portions of the chain completely eliminates sympathetic control of the viscera.

IV

The first fact which strikes the observer is that sympathectomized animals continue to live without apparent difficulty. For about three and a half years we kept in the Laboratory, in good health, an animal (cat 107) which had had no part of the sympathetic system since October 10, 1927. In April, 1931, it was sacrificed for study of its tissues. This study, conducted by Clark, revealed surprisingly few differences between the normal and the sympathectomized animal. The mammary glands showed signs of atrophy, as in senescence; and the thyroid structure had anomalous areas of small follicles with columnar cells and little colloid and other areas of very large irregular follicles with flat epithelium and much colloid. Aside from these changes, which must be cautiously interpreted, the effects of the operation were slight. These facts definitely contravene certain views which have been put forth by previous workers. Meltzer, for example, removed the superior cervical ganglia from rabbits and cats, and because of the very high mortality, he concluded that those ganglia contain a principle which is essential for the maintenance of life. Later Spadolini declared that complete and extensive extirpation of the mesenteric nerves of the cat is incompatible with continued existence. He observed a general depression, grave wasting, the lowering of body temperature, and severe lesions of the gastro-intestinal tract. Although we have removed the abdominal sympathetic chains, including the splanchnic nerves, from a large number of cats, and although we have also taken from some of them the semilunar ganglia and have even stripped the nerve strands from the large branches of the abdominal aorta, we have never seen any serious symptoms as a consequence of the procedure. From these observations we are driven to the conclusion that the superior cervical and the sympathetic ganglia and the sympathetic nerves distributed to the stomach and intestines are not essential for life.

One might suppose that the "vegetative" nervous system—even the sympathetic division—would be important for growth. We have, however, performed one-sided sympathectomy in young kittens and permitted them to grow to adult size. Thereafter we have etherized them until they were dead and compared the weights of various bilaterally symmetrical organs and parts, including bones. There were no demonstrable differences between the two sides. It appears, therefore, that the sympathetic system is not concerned with the growth of the skeleton or of internal organs.

We have made careful observations on the basal metabolism before operation and after removal of each of the three portions—cervical, thoracic and abdominal—of the sympathetic chains. Usually there is a slight fall of the metabolic rate after sympathectomy. In our experience this has been most persistently observed after excision of the cervical portion. As a rule, however, the operation does not reduce the basal metabolic rate more than ten per cent, and the significance of that reduction may be questioned.

The suggestions that possibly the sympathetic controls the tonus of skeletal muscle and that it may be serviceable in resisting fatigue raises a question as to the influence of sympathectomy on muscular function and performance. After a unilateral sympathectomy the knee-jerk reflexes on the two sides of the body are not demonstrably different. We have observed also that after unilateral sympathectomy of the lower abdomen, in a dog trained to run in the treadmill, there was no reduction in the total performance of the animal, nor was there any notable difference of action of the two hind limbs.

In clinical literature there are many references to hypothetical "vagotonic" and "sympathicotonic" states and to "autonomic inbalance." The concept underlying the use of these terms is that in normal circumstances the sympathetic and the cranial divisions of the autonomic system are acting constantly in opposition to each other and that the resultant of the conflict is an equilibrium between the two. There is evidence of such opposition in some organs but not everywhere. Where such opposition does not exist a condition of "autonomic inbalance" could not be expected.

But even where there might be conflict the changes which follow excision of the sympathetic ganglia do not point definitely to an unopposed dominance of the antagonistic system. At first the pupil is constricted, but it becomes later less constricted. At first the denervated blood vessels are dilated, but there is evidence that later they regain to some degree a local tonic state. In short, permanent abnormal conditions are remarkable for their absence.

So much for the failure of sympathectomy to produce striking alterations in the body. Now let us turn our attention to some of the defects.

V

In our first published report we stated that sympathectomy does not prevent the female from performing the functions of reproduction and lactation. The observation on which that report was based was made on an animal sympathectomized only a short time before the birth of her young. Figure 38 is a photograph of this animal with her two kittens. Above the animal are shown her two sympathetic strands. This animal did, indeed, nurse one of her young until it was capable of foraging for itself. Later, however, another female from which the sympathetic had been removed for many months gave birth to three kittens. The extraordinary fact appeared that the breasts were not prepared for lactation. Also a bitch, for some time deprived of her sympathetic nerves, had defective function of the mammary glands. Twice she gave birth to large litters of puppies; and on both occasions, Bright and I noted, the breasts did not enlarge, the areolae remained flat and dry and the nipples were hard, so that the offspring could not obtain adequate nourishment. Bacq, in 1932, noted similar disturbances of the denervated mammary glands of rats. The observations on the cat have been confirmed and extended by Simeone and Ross who, besides noting impairment of lactation when the animals gave birth to young a considerable period after sympathectomy, found that abortions, delayed parturition and the delivery of dead fetuses were common phenomena. Interference with milk secretion and also prolonged labor were reported by Bacq, in 1932, as resulting from removal of the abdominal sympathetic chains from rats.

Although sympathectomy does not prevent the female from producing young, it renders the male sterile. As Bacq showed in 1931, in studies on rats and guinea pigs, the male becomes impotent after sympathetic denervation of the genital organs because thereafter the ejaculatory reflex fails to function. There is evidence that the same defect is produced by that operation in man.

When the sympathectomized cat engages in even moderate muscular activity it is promptly revealed as very defective. The blood pressure, instead of rising, actually falls, as was shown by Freeman and Rosenblueth; the animal is therefore incapable of prolonged or vigorous action. The fall of arterial pressure is apparently due to vasodilator impulses carried to the blood vessels over non-sympathetic channels, for Rosenblueth and I found that when we stimulated depressor points in the medulla of the sympathectomized cat the blood pressure dropped. Furthermore, Bacq, Brouha and Heymans recorded a drop of pressure, in the cat without sympathetic system, when they stimulated the nerves of the carotid sinus. In the dog conditions are markedly different. Unlike the cat the dog has only few non-sympathetic vasodilator nerves and is well supplied with sympathetic dilators; stimulation of the carotid sinus nerves in the sympathectomized dog does not lower blood pressure. In consequence the dog, deprived of the sympathetic system, may have no fall of pressure on engaging in activity, but instead, a rise. Time is required for adjustment after the operation, for, as Freeman has noted, even the dog may faint on exertion in the early stages of recovery. When fully recovered, however, Brouha, Dill and I found that the dog deprived of sympathetic nerves is capable of running, playing, jumping and fighting almost as efficiently as the normal animal. Jourdan and Nowak's demonstration of the presence of cardio-accelerator fibers in the vagus nerves of the dog, capable of raising the heart rate to a greater degree than is reached by mere abolition of vagal tone, offers further explanation of the remarkable performance of the dog, as contrasted with the cat, surviving complete sympathectomy.

Because the cat is not easily trained to exercise, a quantitative comparison of the ability of the dog and the cat to work is not easy. Many of the physiological devices which are used in running, however, are brought into action when animals are exposed to a low concentration of oxygen or to a high altitude. In their endurance of low oxygen tensions it is readily possible to compare the dog and the cat, both normal and sympathectomized. When such tests are applied a striking difference is found between the two species. Sawyer, Schlossberg and Bright exposed normal cats for an hour to an atmosphere containing 8 per cent oxygen (equivalent to an altitude of about 24,000 feet) and noted that the animals were little disturbed; sympathectomized çats, on the other hand, when placed in the same conditions, fainted or collapsed in a relatively short time, occasionally in 15 or 20 minutes. McDonough, using similar methods, found that normal dogs would live in an atmosphere of approximately 4 per cent oxygen (representing an altitude considerably higher than Mt. Everest) more than two hours before manifesting, by a change to slow breathing, that they were in danger. Sympathectomized dogs are not so resistant; in six cases this low oxygen concentration was endured for an hour and a half by four of the animals, and about two hours by the other two. Clearly the homeostatic mechanisms are more effective in the dog than in the cat, but in both animals removal of sympathetic control reduces the ability to meet stress.

The homeostasis of blood sugar, also, is markedly influenced by sympathectomy. This fact was established by determining the effect of emotional excitement on blood sugar before and after sympathetic extirpation. As shown in figure 39 excitement from three to ten minutes causes in the normal cat an average increase of blood sugar amounting to 34 per cent. In sympathectomized cats, on the other hand, similar durations of excitement result in no consistent rise in the glycemic level. Corroborative evidence is obtained by giving insulin. We have already seen that denervation of the adrenal glands renders cats much more sensitive to insulin hypoglycemia than are normal animals (see p. 113). Removal of the sympathetic system not only abolishes nervous control of adrenal secretion, but also, as Bodo and Benaglia have shown, it excludes the possibility

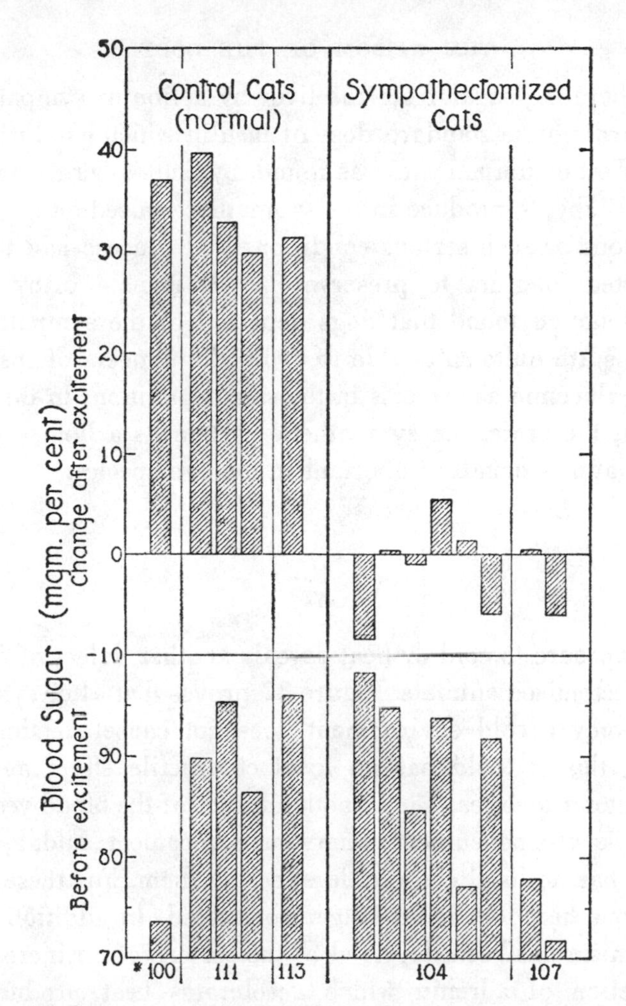

Fig. 39. Increases of blood sugar in normal cats after emotional excitement and absence of any consistent effects on blood sugar in sympathectomized cats that were similarly excited. The numbers designating the cats are given below.

of liberating sugar from the liver by action of sympathin. Accordingly, a standard dose of insulin which has little or no effect on normal cats was found, by Schlossberg, Sawyer and Bixby, to produce in the sympathectomized either convulsions or such serious conditions that glucose had to be injected in order to preserve life. It is noteworthy that McDonough found that dogs deprived of the sympathetic chains are quite as unable to avoid the dangers of insulin hypoglycemia as are cats in the same condition. In this respect, therefore, the sympathetic system as a homeostatic regulator is equally important in the two species.

VI

Exposure to cold or heat reveals another defect of sympathectomized animals. Figure 37 proves that after sympathectomy a cold environment does not cause erection of hair, that is, cold has no local contractile effect on the pilomotor muscles. The smooth muscles of the blood vessels are likewise not subject to nervous government; and, again, cold has no local contractile effect on them. For these two reasons heat loss is no longer controlled. In addition, the organism has been deprived of the service of an increased secretion of adrenin which accelerates heat production when the body temperature tends to fall. For these various reasons the sympathectomized animal, when confronted by the problem of maintaining its normal temperature, is physiologically defective. The behavior of such an animal in cold weather is consistent with this defective state. It exhibits a marked antipathy to cold air and to drafts. In the cold weather of winter it crouches near sources of heat and leaves such places only at feeding time.

To test sympathectomized animals regarding their reaction to cold we have placed them in a refrigerating room, the temperature of which ranged between 0.8° and 6.0° C. (33.4° and 42.8° F.), and have made observations on the body temperature and on shivering. Results with cats are shown in figure 40. The solid lines trace the temperature changes of normal animals during their stay in a cold environment for about two hours. The dash lines represent the changes in sympathectomized animals under similar conditions. Note that observations were made on cats 393 and 396 both before and after sympathectomy. As the records for 393 clearly prove, there is a striking difference between the response of the animal in the two conditions. After sympathectomy there was a much deeper depression of the body temperature—indeed, there was a rapid drop of temperature to a low point at which it could be fairly maintained. Also there was a much greater frequency and prominence of shivering. This response, which is indicated by zigzag lines, did not appear at all in cat 393 before sympathectomy but was quite marked after that operation. The phenomenon was very prominent in the other animals which had been deprived of their sympathetic systems. Since shivering was the only resource, except exercise, that was left to protect the animals from a deep drop of temperature, the actual low temperature which was maintained, as, for example, in cat 400, probably indicates an equilibrium between the heat lost to the cold surroundings and the heat produced by the shivering process. Observe especially the record of cat 107. This was the animal which I mentioned earlier as having had no sympathetic system after October 10, 1927. During the year that had elapsed since that date the animal appears to have become accustomed to reliance upon shivering as the main protection against a temperature fall. On being placed in the cold room it very soon began to shiver vigorously and by so doing prevented a marked drop in temperature and indeed managed to maintain its normal temperature level.

When dogs, normal and sympathectomized but of similar size and hair coat, were compared in a cold room (about 5° C.), McDonough observed vigorous shivering in all instances. In normal dogs the shivering rate was recorded at about 400 muscular contractions per minute; in the sympathectomized it was about 600. Despite the higher rate, however, there was a fall of body temperature. It was not so great a fall as that observed in cats deprived of the sympathetic system, being only about 0.5° C., but it was typically present. From these observations we may conclude that because of its service in lifting the hair, in contracting surface blood vessels and discharging adrenin, the sympathetic system is useful to dogs as it is to cats in maintaining a stable state. When the system is absent, shivering occurs sooner and is more intense than in normal animals. The dog, however, differing from the cat in being much more muscular, produces more heat in the shivering process. Furthermore, even in his normal state he is accustomed to shivering as a means of preserving temperature homeostasis. These two conditions, I suggest, explain the greater effectiveness of the dog as compared with the cat when they are exposed to cold after the sympathetic system has been removed.

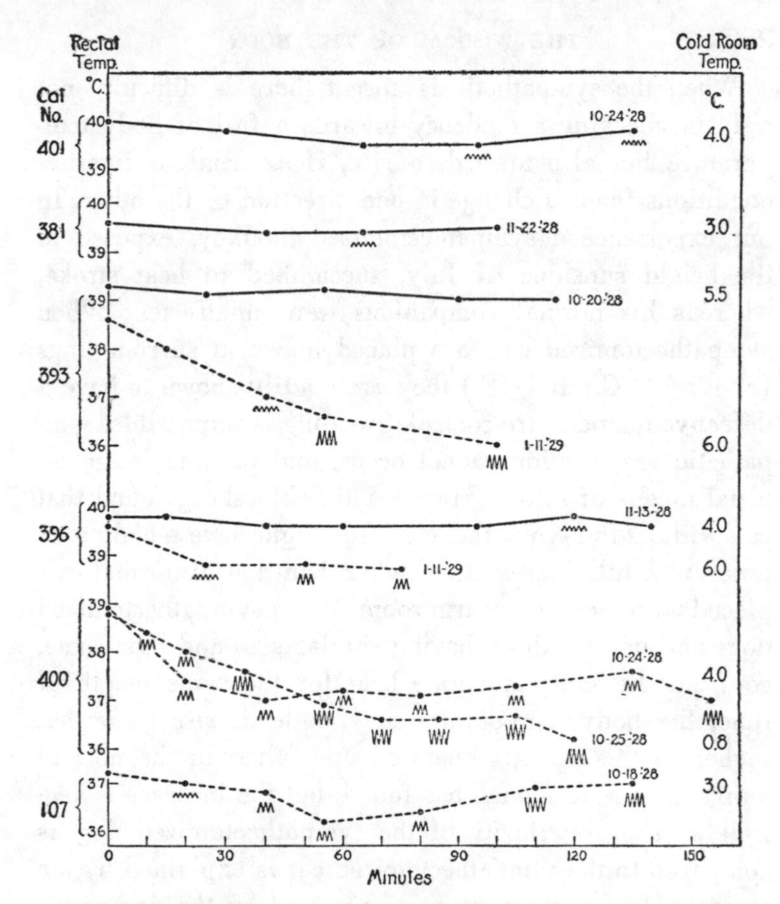

Fig. 40. Records of the bodily temperature of cats kept in a cold room. Continuous lines show the effects on normal cats; dash lines the effects on sympathectomized cats. Shivering is represented by zig-zags under the record; the size of the zig-zags indicates roughly the intensity of the shivering reaction.

When the sympathetic is absent there is difficulty not only in checking a tendency towards a fall of body temperature but also towards a rise, if external or internal conditions favor a change in one direction or the other. In our experience a sympathectomized monkey, exposed to the bright sunshine of July, succumbed to heat stroke, whereas his normal companions were unaffected. When sympathectomized cats are placed in warm surroundings (about 40° C.; 104° F.) they are readily shown to have a defective temperature control. Sweating is impossible, sympathetic vasodilation cannot occur, and panting is an unusual means of relief. Sawyer and Schlossberg found that cats without the sympathetic system might have a body temperature a full degree (C.) higher than that of normal cats placed with them in a warm room. When sympathectomized dogs and normal dogs, having similar size and hair coats, confront the same moderate heat for the same length of time, the body temperature may, indeed, rise somewhat higher in the animals operated upon than in the normal animals, as McDonough has found, but the increase is negligible. The superiority of the sympathectomized dog as compared to the sympathectomized cat is explained by the remarkable accessory devices possessed by the dog in his great tongue and lavish discharge of saliva. When his temperature tends to rise he promptly resorts to panting. By the evaporation of saliva, heat is removed from the tongue and mouth where the blood circulates abundantly, and thus the body temperature is kept from rising.

Still another noteworthy difference between normal and sympathectomized animals is seen when hemorrhage occurs. We have previously noted that then the sympathetic system becomes active and, by inducing vasoconstriction and discharging red corpuscles from the spleen, compensates for the blood loss (see pp. 51–54). Schlossberg and Sawyer confirmed the observation, recorded in figure 9 (p. 53), that from 30 to 40 per cent of the blood can be taken from normal cats in three or four bleedings before the compensatory vasoconstrictor mechanism fails to raise the reduced blood pressure. They found that sympathectomized animals are much more sensitive to hemorrhage; in them the compensatory reaction is in some cases absent and in others very slight after a single removal of as little as 13 to 15 per cent of the total blood volume. In the absence of the sympathetic system the mechanisms for preserving an effective fluid matrix are largely ineffective.

VII

The results which I have just reported may seem more impressive because of the slight effects resulting from removal of the sympathetic system than because of actual deficiency phenomena. It must be admitted, however, that the slight effects may be misleading. The animals, to be sure, continue to live, but they live in the protective confines of the laboratory where there are no marked temperature changes throughout the year and no necessity for struggle for food, no requirement of escape from enemies, no danger of hemorrhage. This is a very special and limited sort of existence. Judging by such observations we might easily draw the inference that the sympathetic system is of minor importance for the proper functioning of the body. Such an inference would be erroneous.

The dog, admittedly, is less affected than is the cat by loss of sympathetic control of his homeostatic mechanisms. He withstands heat, cold, muscular work and lack of oxygen much better than the sympathectomized cat. These facts do not prove, however, that the sympathetic system is unimportant for the maintenance of homeostasis. They emphasize the point that the dog has remarkable accessory functions for that purpose. They are associated, as I interpret them, with the extraordinary activity of the dog. He is naturally a very energetic, running animal. As such he produces much heat. Lacking abundant sweat glands he has the special ability to discharge heat by panting—an activity seldom seen in cats. In resisting cold, also, the dog is exceptional in shivering readily and because of his powerful muscles producing a large amount of heat. Furthermore, as a running animal the dog requires a plenteous supply of oxygen to burn the lactic acid resulting from his vigorous muscular activities. All the physiological devices for supplying oxygen to the tissues are much more developed in the dog than in the cat. His lung volume, his cardiac capacity, his blood weight in relation to body size, and also his hemoglobin percentage, are all remarkably greater than in the cat. Note that none of these advantages is reduced by sympathectomy. In this group of functions concerned with delivery of oxygen to active organs the heart rate alone might be affected by sympathectomy; but the dog has vagal cardio-accelerators which can replace, in part, the absent sympathetic accelerators. After extirpation of the sympathetic system these adjuvant homeostatic devices come prominently into play; and because the dog has them in much greater measure than the cat he can meet stresses much more effectively, though the regulatory nerves are lacking.

If sympathectomized animals were set free in the outer world and had to meet its demands in struggle for food, safety and warmth, they would be found more or less defective according to the variable efficiency of their accessory stabilizing mechanisms. Even in the favorable conditions displayed by the sympathectomized dog, however, absence of sympathetic control of corrective devices is accompanied by an inability to preserve constancy of the internal environment though the stress is only moderate.

References

Bacq. Am. Journ. Physiol., 1931, xcvi, 321.
Bacq. Ibid., 1932, xcix, 444.
Bacq, Brouha and Heymans. Arch. internat. de Pharmacodyn. et de Thérap., 1934, xlviii, 429.
Bodo and Benaglia. Am. Journ. Physiol., 1938, cxxi, 728.
Brouha, Cannon and Dill. Journ. Physiol., 1936, lxxxvii, 345.
Cannon and Bright. Am. Journ. Physiol., 1931, xcvii, 319.
Cannon, Newton, Bright, Menkin and Moore. Ibid., 1929, lxxxix, 84.
Clark. Journ. Comp. Neurol., 1933, lviii, 553.
Freeman. Personal communication.
Freeman and Rosenblueth. Am. Journ. Physiol., 1931, xcviii, 455.
Izquierdo and Cannon. Ibid., 1928, lxxxiv, 545.
Jourdan and Nowak. C. r. Soc. de Biol., 1934, cxvii, 234.
McDonough. Am. Journ. Physiol., 1939, cxxv, 530.
Meltzer. Proc. Nat. Acad. Sci., 1920, vi, 532.
Menkin. Am. Journ. Physiol., 1929, lxxxv, 489.
Rosenblueth and Cannon. Ibid., 1934, cviii, 599.
Sawyer and Schlossberg. Ibid., 1933, civ, 172, 195.
Sawyer, Schlossberg and Bright. Ibid., 1933, civ, 184.
Schlossberg, Sawyer and Bixby. Ibid., 1933, civ, 190.
Simeone and Ross. Ibid., 1938, cxxii, 659.
Spadolini. Arch. Ital. de Biol., 1926, lxxvii, 17.

XVIII

THE GENERAL FEATURES OF BODILY STABILIZATION

I

IN THE previous chapters we have witnessed an inductive unfolding of the methods employed in preserving homeostatic conditions and we have examined the evidence that these conditions are under the government of the interofective, autonomic nervous system. A review of the facts, with the purpose of drawing from them the general principles which they illustrate, will be useful in giving us an opportunity to look at them from a new point of view and also in preparing us for an inquiry into the possibly necessary prerequisites for stability in other types of organization.

One of the most striking features of our bodily structure and chemical composition that may reasonably be emphasized, it will be recalled, is extreme natural instability. Only a brief lapse in the coördinating functions of the circulatory apparatus, and a part of the organic fabric may break down so completely as to endanger the existence of the entire bodily edifice. In many illustrations we have noted the frequency of such contingencies, and we have noted also how infrequently they bring on the possible dire results. As a rule, whenever conditions are such as to affect the organism harmfully, factors appear within the organism itself that protect it or restore its disturbed balance. The *types* of arrangement by which this stabilization is accomplished is our present interest.

Although some organs are under a sort of control that keeps them from going too fast or too slowly—the heart with its inhibitor and accelerator nerves is an example—these instances may be regarded as secondary and contributory forms of self-regulation. In the main, stable states for all parts of the organism are achieved by keeping uniform the natural surroundings of these parts, their internal environment or fluid matrix. That is the common *intermedium* which, as a means of exchange of materials, as a ready carrier of supplies and waste, and as an equalizer of temperature, provides the fundamental conditions which facilitate stabilization in the several parts. This "milieu interne," as Claude Bernard pointed out, is the product of the organism itself. So long as it is kept uniform, a large number of special devices for maintaining constancy in the workings of the various organs of the body are unnecessary. The steadiness of the "milieu interne," therefore, may be regarded as an arrangement of economy. And the course of evolution of higher organisms has been characterized by a gradually increasing control of the functions of that *milieu* as an environmental and conditioning agency. Just insofar as this control has been perfected, both internal and external limitations on freedom of action have been removed and risks of serious damage or death have been minimized. The central problem in understanding the nature of the remarkable stability of our bodies, therefore, is that of knowing how the uniformity of the fluid matrix is preserved.

II

A noteworthy prime assurance against extensive shifts in the status of the fluid matrix is the provision of sensitive automatic indicators or sentinels, the function of which is to set corrective processes in motion at the very beginning of a disturbance. If water is needed, the mechanism of thirst warns us before any change in the blood has occurred, and we respond by drinking. If the blood pressure falls and the necessary oxygen supply is jeopardized, delicate nerve endings in the carotid sinus send messages to the vasomotor center and the pressure is raised. If by vigorous muscular movements blood is returned to the heart in great volume, so that cardiac action might be embarrassed and the circulation checked, again delicate nerve endings are affected and a call goes from the right auricle, that results in speeding up the heart rate and thereby hastening the blood flow. If the hydrogen-ion concentration in the blood is altered ever so slightly towards the acid direction, the especially sensitized part of the nervous system which controls breathing is at once made active and by increased ventilation of the lungs carbonic acid is pumped out until the normal state is restored.

In previous pages we have learned of other instances in which quick and efficient corrective measures are instituted when the first intimations of disturbance appear, without our having, however, clear knowledge of what the indicator is or how it works. The operation of the sympathico-adrenal apparatus to increase blood sugar when the glycemic percentage begins to fall below the critical level is a case in point. What sets it going we do not know. And the regulation of body temperature probably belongs in this class. Although the diencephalon seems to contain the thermostat, it may be that the controlling center is there and that it is managed by an agent outside—a number of recent discoveries in physiology have deprived the brain of credit for direct action and have proved that the true *modus operandi* is a reflex. Here more information is required before we can be sure of the location of the agency which is on guard.

Indicators of variation in still other states in the fluid matrix, which are regularly held in remarkable steadiness and which, if altered, are soon restored, are unfortunately unknown. The blood proteins (on which the very existence of the normal blood volume depends), the blood calcium (of primary importance for the proper functioning of the neuromuscular system) and the red corpuscles of the blood (essential for the oxygen supply to the tissues) are examples of factors in the fluid matrix, all of which exhibit homeostasis to a surprising degree. Marked change in their concentration brings about alarming disturbances in the organism. In all probability a slight movement in the direction of change is signalled, just as in other instances which we have considered above, and then the tendency is corrected. But what influences the signal and how the signal sends orders to the organs which make the correction, must remain a mystery until further physiological research has disclosed the facts.

III

Two general types of homeostatic regulation can be distinguished, dependent on whether the steady state involves *materials* or *processes*. We shall consider first the regulation of materials.

The homeostasis of materials, as numerous instances have shown us, is accomplished by *storage* as an adjustment between occasional or precarious supply and constant, and at times augmented, need. Storage, we have seen, is of two kinds—*temporary*, for immediate accommodation and use, and *reserved*, for later and lasting service.

The temporary storage is apparently a consequence of simple flooding of tissue spaces. When an abundance is provided it spreads into the fine meshwork, the "spongy cobweb of delicate filaments," of which the connective tissue under the skin and around and between the muscle bundles is largely composed. In this region water is stored and all constituents of the blood which are dissolved in water—salts especially, and sugar. As the high levels which produced the flooding subside, the substances seep back into the running stream (the blood), by which they are distributed to other parts where need may exist. Or, in the case of glucose, the temporary storage may be changed to the permanent form, without immediate use. This simple mode of setting aside materials we have called *storage by inundation*. It appears to have no specially developed control other than the relative concentration of the substances concerned, either in the blood or in the fluids of the alveolar connective tissue.

In the more or less permanent or reserved storage the

materials are set apart inside of cells or in special places. It has been called *storage by segregation.* From what we know regarding the management of this mode of storage in certain cases, we may infer that it differs from storage by inundation in being subject to a nervous or neuro-humoral government. The best understood instance of such government is that of glucose regulation in the blood. We have seen evidence that when the blood sugar rises much above the usual level of 100 milligrams per cent, the vago-insular mechanism is stimulated, and the insulin discharged into the blood stream from the island cells of the pancreas limits the rise by favoring the use of the sugar and its storage in the cells of the liver and of muscles. On the other hand, if the glycemic percentage falls much below the usual level, a critical point is reached at which the sympathico-adrenal mechanism is brought into action. The glycogen stored in the liver cells is thereby transformed into glucose. This escapes through the cell walls into the circulating blood, and thus the tendency towards a low blood sugar and its harmful consequences is averted.

It must be admitted that, in relation to the homeostasis of other materials, our acquaintance with its regulation is largely restricted to suggestive indications. We have learned that calcium is set aside as a reserve in the spicules and trabeculae inside the long bones. Its abundance in those forms when the calcium of the food is plentiful, and its disappearance when the intake is meager, clearly prove the fact of storage. Furthermore, some evidence at hand, which points to the functioning of the parathyroid glands as influential in promoting storage and to the thyroid as a releasing agent, might offer a fairly close analogy to the regulation of blood sugar. But the relations between the glands and the hoarding or liberating of calcium are at present too loosely defined to permit any reliable conclusion to be drawn. And we are still in the dark as to factors which excite or check the activity of the glands!

The same sort of suggestive evidence can be cited with regard to the regulation of other materials. Fat and protein, like calcium, are stored by segregation; the fat inside the cells of adipose tissue, the protein, according to present testimony, inside of liver cells. We have noted that one observer has reported that the protein masses in the liver cells disappear if adrenin is injected into the circulation. If that observation is confirmed, it may lead to further insight into a sympathico-adrenal influence on the protein content of the blood. Further, we know that when the thyroid gland is deficient and also when the part of the base of the brain near the pituitary gland has been injured, a heavy layer of fat is developed under the skin and elsewhere in the body. Also we know that the fat which is stored because of thyroid deficiency is readily removed by feeding thyroid gland or an extract of it. All this information hints at interesting possibilities of regulation of the supplies of both protein and fat. The importance of securing further knowledge of the ways in which these primary substances are laid by and later mobilized for use is unquestioned. But for such knowledge we must await the progress of patient research.

Back of storage, and, indeed, as occasion for storage, are the motivators for the taking of food and water. Fundamentally these are the disagreeable experiences of hunger and thirst: the unpleasant pangs, which disappear when food is eaten; and the unpleasant dryness of the mouth, that disappears when water or watery fluid is drunk. But these automatic "drives" lead at times to delectable sensations of taste and smell. Such sensations become associated with the taking of the special foods and drinks which have occasioned them. Thus appetites are established which, by *inviting* to eat and to drink, may replace in part the need for the goads of hunger and thirst. But if appetites fail to keep up the supplies, the more imperious and more insistent agencies come into action and demand that the reserves be replenished.

IV

Another means of assuring constancy of the fluid matrix is by *overflow.* This arrangement sets a limit on the upward variation of substances in the fluid matrix. Already, in relation to the homeostasis of glucose, we have noted the use of overflow as a means of checking too great a rise of that constituent of the blood. Not only excess of sugar, however, but also excess of water and of certain substances dissolved in it—sodium and chloride ions, for example—is discharged by way of the kidneys. In accordance with the modern theory of urine formation these are all "threshold substances"; they are reabsorbed by the kidney tubules only in such quantitative relations to one another as to preserve the normal status of the blood. All above these amounts is allowed to escape from the body as overflow.

It is interesting to note that the threshold substances are primarily stored by flooding or inundation. When a sufficient reserve of these supplies has been established, however, the ability of the overflow mechanism to maintain homeostasis is marvelous. I have previously called attention to the experiments of Haldane and Priestley on themselves in which during six hours an amount of water exceeding by one-third the estimated volume of the blood was allowed to overflow through the kidneys with such delicacy that at no time was the blood diluted to a degree which notably reduced the percentage of hemoglobin. These experiments demonstrate vividly not only the efficacy of the kidney as a spillway but also the use of the principle of the spillway as a means of maintaining a uniform state in the fluid matrix.

In their functions as overflow organs the kidneys act to preserve the normal balance between acid and base in the blood. If non-volatile acid is produced in excess, it passes the barrier; and if there is too much alkali in the blood, it also overflows and escapes.

Not only the kidneys but also the lungs serve for overflow. As we have seen, a very slight excess of carbonic acid in the arterial blood instantly induces deeper breathing. The increased pulmonary ventilation, occasioned thereby, promptly and effectively reduces the carbon dioxide in the alveoli, so that, in spite of large production of carbonic acid, the percentage of the gas in the alveolar air is kept nearly constant. By this means provision is made for the extra carbon dioxide to stream out from the blood over a dam set at a fixed level. In consequence, under usual conditions, the hydrogen-ion concentration of the blood is evenly maintained, and the harmful effects of an excessive shift in the acid direction is avoided.

It is noteworthy that overflow is used as a regulatory

process not only for keeping down the concentration of waste material (carbon dioxide), but also for keeping down the concentration of useful material (glucose). Here again we see indicated the importance of constancy of the fluid matrix as a primary condition of the organism.

V

The second general type of homeostatic regulation is that in which *processes*, rather than materials, are involved. The most significant instance is presented in the mechanism of heat regulation. The processes of heat production and heat loss, as we know, are going on continuously. When the body temperature starts to fall the process of heat production is accelerated and that of heat loss is diminished. And when the opposite tendency of the body temperature appears, the effects on the processes are reversed. Thus by altering the rate of continuous processes, in a manner nicely adapted to the needs of the organism, the temperature is held in a remarkably uniform balance.

Similar phenomena are found in the supply of oxygen to the tissues, and in the preservation of the normal acid-base equilibrium of the blood. Continuously oxygen is being supplied by moderate activity of the respiratory and circulatory systems, but in special circumstances the demand is greatly augmented. Thereupon, the respiration rate and the circulation rate are correspondingly accelerated, the red blood corpuscles are loaded and unloaded much faster, and the oxygen tension in the fluid matrix near the needy organs, in spite of their larger requirement, is to a high degree sustained. As the special circumstances which have roused the systems to extra action fade away, the systems, after paying any "oxygen debt" that may be left, subside into their former moderate functional speed. The same principles and many of the same phenomena are illustrated when conditions cause an increase of acid concentration in the blood. And when the reaction of the blood is moved towards the alkaline side, again a continuous process, that of respiration, is retarded or wholly stopped until enough of the continuously produced acid has accumulated to restore the normal acid-base relation.

A combination of the use of reserves and the use of the altered rate of processes is found in the complex of mechanisms which operate to assure uniformity of the oxygen tension. It will be recalled that besides the faster return of blood to the heart, the faster heart beat, and the high head of arterial pressure which results in a faster blood flow—all accelerations of continuous performances—there is a setting free of the concentrated red corpuscles from the store in the spleen. These corpuscles now join those already in the hurried current and help them to meet the exigency which confronts the laboring cells.

VI

The modes of storage and release, and the speeding up and slowing down of continuous processes, which keep steady the conditions of the internal environment, are not, as a rule, under control from the cortex. We can, to be sure, voluntarily breathe faster or slower, but ordinarily the rate of respiration is managed automatically. And in a like manner all the other homeostatic adjustments are managed.

Automatically materials are set aside in the reserves by the natural functioning of the cranial division of the interofective nervous system. Automatically the reserves are called forth and the processes are accelerated when the blood sugar runs low, when extra oxygen is needed, when acid tends to accumulate or when the temperature begins to fall. It is a special part of the autonomic nervous system —the sympathico-adrenal division—which is charged with the performance of these services, quite outside of conscious direction; indeed, in ways which only elaborate physiological research has revealed.

Whether the sympathico-adrenal division is aroused by pain or excitement, by muscular effort, asphyxia or low blood pressure, by cold or hypoglycemia, the presenting situation is one in which the constancy of the fluid matrix is endangered or is likely to be endangered. In each one of these contingencies the operation of the system is such as to favor the welfare of the organism by preserving homeostasis of the internal environment. The blood flow is shifted and hastened so as to maintain uniformity of the oxygen tension and to keep level the acid-base balance during muscular exertion; the metabolic rate is raised so as to produce more heat if the heat loss is excessive; glucose is liberated from hepatic stores when the amount in the blood is dropping to a low percentage or when there may be special need for it; the capacity of the vascular system is adapted to a reduced blood volume when, after hemorrhage, the circulating blood as a common carrier is likely to become inefficient—in short, as these illustrations indicate, the sympathico-adrenal apparatus promptly and automatically makes the adjustments which are required to preserve the normal internal condition for the living parts when that is disturbed or likely to be disturbed.

The amazing feature of the rôle played by the sympathico-adrenal system is its applicability to the wide range of possible disturbances that we have just noted. As stated earlier, the system commonly works as a unit. It is very remarkable indeed that such unified action can be useful in circumstances so diverse as low blood sugar, low blood pressure, and low temperature. It would seem, however, that the utility is not always complete in details, for at times effects are produced which apparently have little or no value for the organism. Examples of these meaningless responses are seen in the sweating in hypoglycemia and in the rise of blood sugar in asphyxia. In other circumstances, as we well know, these are serviceable responses; e.g., sweating when vigorous muscular exercise produces extra heat, and liberation of sugar from the liver when the glycemic percentage falls too low. The appearance of inappropriate features in the total complex of sympathico-adrenal function is made reasonable, as I pointed out in 1928, if we consider, first, that it is, on the whole, a unitary system; second, that it is capable of producing effects in many different organs; and third, that among these effects are different combinations which are of the utmost utility in correspondingly different conditions of need. Vasoconstriction, increased heart rate, and a larger output of adrenin are a useful group of sympathico-adrenal reactions when the blood pressure is low or when the temperature must be sustained; and that group is not any less effective in the two situations because there is an attendant, perhaps useless, increase of blood sugar. The effects which, in any particular case of need, are not useful may reasonably be regarded as incidental, as lying outside the group of sympathico-adrenal agencies which, for the moment, are working for homeostasis.

VII

In 1926 I advanced a number of tentative propositions concerned with steady states in the body, and with the maintenance of these states, that are pertinent in the present consideration of the general features of homeostasis. We shall now regard only four of them.

"In an open system, such as our bodies represent, compounded of unstable material and subjected continually to disturbing conditions, constancy is in itself evidence that agencies are acting, or ready to act, to maintain this constancy." This is an inference based on insight into the ways by which some steady states are regulated (e.g., glycemia, body temperature, and the acid-base balance), and it was expressed with confidence that other steady states, not yet fully understood, are similarly regulated. When we have learned more concerning the factors which govern the constancy of proteins, fat and calcium in the blood plasma, we shall probably see that it results from as nice devices as those operating in the better known cases of homeostasis.

"If a state remains steady it does so because any tendency towards change is automatically met by increased effectiveness of the factor or factors which resist the change." Thirst, the reaction to low blood sugar, the respiratory and circulatory responses to a blood shift towards acidity, the augmented processes of heat conservation and production, all become more intense as the disturbance of homeostasis is more pronounced, and they all subside quickly when the disturbance is relieved. Similar conditions probably prevail in other steady states.

"The regulating system which determines a homeostatic state may comprise a number of coöperating factors brought into action at the same time or successively." This statement is well illustrated by the elaborate and complex reactions in the blood itself, and simultaneously in the circulatory and respiratory system, that preserve the relative constancy of the acid-base relation in the plasma; and also by the arrangements for protection against a fall of temperature in which the defensive processes are awakened in series, one after another.

"When a factor is known which can shift a homeostatic state in one direction it is reasonable to look for automatic control of that factor, or for a factor or factors having an opposing effect." This postulate is really implied in the previous postulates and may be regarded as emphasizing the confident opinion that homeostasis is not accidental, but is a result of organized self-government, and that search for the governing agencies will be rewarded by their discovery.

VIII

It is not supposed that the full display of homeostatic adjustments will be found in all forms of animals. The illustrations which I have given in earlier chapters have been taken from experiments and observations on mammals. Birds alone can share with mammalian forms the possession of complicated mechanisms for keeping constant the fluid matrix of the body—and birds have been very little studied with reference to these mechanisms. Reptiles and amphibia are much less highly organized, in the sense of having a controlled internal environment which liberates them from the vicissitudes of the external environment. As pointed out in an earlier chapter, the amphibian is unable to preserve his water content and cannot hold steady his own temperature, independent of that of his outer world. The reptile, a higher type, does not lose water quickly to the air about him as the amphibian does, but, like the amphibian, the reptile is "cold-blooded" and therefore is limited in his activities by cold surroundings.

The evidence that homeostasis as seen in mammals is the product of an evolutionary process—that only gradually in the evolution of vertebrates has stability of the fluid matrix of the body been acquired—is interestingly paralleled in the development of the individual. Indeed, a suggestive addition to the group of facts which support the idea that the history of the individual summarizes the history of the race, or that ontogeny recapitulates phylogeny, is found in the absence or deficiency of homeostatic regulation in babies during a considerable period after birth, and the later rather slow acquirement of control. Before birth, of course, the baby has the benefit of the uniformity of the mother's "milieu interne." At birth he is suddenly exposed to very different and quite variable surroundings, when his own "milieu interne," though formed, has not been subjected to any stress which might alter it. Now it has long been known that the new born, when they are exposed to cold, have little ability to maintain a constant temperature. Instead of reacting swiftly in such ways as would keep the temperature from falling, the organism permits the fall to occur, much as it occurs in cold-blooded forms, without a quiver of resistance. The elaborate complex of adjustments for the homeostasis of temperature, that is characteristic of adults, is only gradually developed, perhaps as the consequence of exercise and training. The control of blood sugar is similarly the result of a developmental process. Recent observations by Schretter and Nevinny have shown that in the early days of babyhood the percentage of glucose in the blood varies much more, and in larger oscillations, than it does during the later periods of life. It seems not improbable that study of other homeostatic regulations would prove that they too are unstable at the start and only by experience acquire the efficiency seen in adults.

IX

Repeatedly in foregoing chapters I have called attention to the fact that insofar as our internal environment is kept constant we are freed from the limitations imposed by both internal and external agencies or conditions that could be disturbing. The pertinent question has been asked by Barcroft, freedom for what? It is chiefly freedom for the activity of the higher levels of the nervous system and the muscles which they govern. By means of the cerebral cortex we have all our intelligent relations to the world about us. By means of it we analyze experience, we move from place to place, we build airplanes and temples, we paint pictures and write poetry, or we carry on scientific researches and make inventions, we recognize and converse with friends, educate the young, express our sympathy, tell our love—indeed, by means of it we conduct ourselves as human beings. The alternative to this freedom would be either submission to the checks and hindrances which external cold or internal heat or disturbance of any other constants of the fluid matrix would impose upon us; or, on the other hand, such conscious attention to storage of materials and to altering the rate of bodily processes, in order to preserve constancy, that time for other affairs would be lacking. It would be like limiting social activities because of domestic duties, or excluding foreign relations because of troubles in the interior. The full development and ample expression of the living organism are impossible in those circumstances. They are made possible by such automatic regulation of the routine necessities that the functions of the brain which subserve intelligence and imagination, insight and manual skill, are set free for the use of these higher services.

In summary, then, we find the organism liberated for its more complicated and socially important tasks because it lives in a fluid matrix, which is automatically kept in a constant condition. If changes threaten, indicators at once signal the danger, and corrective agencies promptly prevent the disturbance or restore the normal when it has been disturbed. The corrective agencies act, in the main, through a special portion of the nervous system which functions as a regulatory mechanism. For this regulation it employs, first, storage of materials as a means of adjustment between supply and demand, and, second, altered rates of continuous processes in the body. These devices for maintaining constancy in the organism are the result of myriads of generations of experience, and they succeed for long periods in preserving a remarkable degree of stability in the highly unstable substance of which we are composed.

References

Cannon. Ergebn. d. Physiol., 1928, xxvii, 380.
Cannon. Am. Journ. Med. Sci., 1926, clxxi, 1.
Barcroft. Journ. Exper. Biol., 1932, ix, 24.
Schretter and Nevinny. Zeitschr. f. Geburtsh. u. Gynäkol., 1931, xcviii, 258.

EPILOGUE

RELATIONS OF BIOLOGICAL AND SOCIAL HOMEOSTASIS

I

ARE there not general principles of stabilization? May not the devices developed in the animal organism for preserving steady states illustrate methods which are used, or which could be used, elsewhere? Would not a comparative study of stabilizing processes be suggestive? Might it not be useful to examine other forms of organization—industrial, domestic or social—in the light of the organization of the body?

These are tempting questions. Many times in the history of philosophy and sociology similar questions have led to an examination of the analogies between the body physiologic and the body politic. The biologist is as subject to temptation in respect to these analogies as are the philosophers and sociologists! He may lack the philosophers' broad outlook and the sociologists' knowledge of the complex details of the social system. But as a unit of that system he is interested in it. And he looks on the analogies from the biological point of view. May not the new insight into the devices for stabilizing the human organism, which we have been examining in the foregoing chapters, offer new insight into defects of social organization and into possible modes of dealing with them? The details of bodily homeostasis are, of course, available to anyone who cares to see whether they offer any suggestions for the study of social conditions. As a stimulus to such suggestions it might not be amiss to consider some features of their apparent analogies.

II

In an earlier chapter I pointed out that the single-cell organism, living in a flowing stream, is dependent immediately on its surroundings; it has no means of controlling the environment and must submit wholly to what the environment imposes upon it. Only when cells grow in masses do they acquire the possibility of developing an internal organization, capable of separating them from the disturbances due to shifts of external circumstance.

We must not overlook the fact that when cells grow in masses they still remain living units. Like the isolated single cell each cell in a complex organism has its own life processes. In our discussion of homeostasis we have considered the environment—the internal environment—which is provided for these living units. We have not, however, regarded the events occurring within the units themselves. Each one takes in, from its fluid contact, water and salts and oxygen; it takes in food which it uses to build up or repair its own structure, or to elaborate new substances for special secretions, or to secure the energy needed in performing other special services for the organism as a whole; and finally it discharges the waste resulting from the wear and tear and from the débris of its own activities. All these complicated functions the cell normally carries on in a nicely adjusted manner, with not too much and not too little of either intake or output. And throughout the multitudes of exchanges which are involved it preserves in a marvelous manner its intimate texture and precise action.

In the one-cell organism all the vital functions—digestion, motion, reproduction—are performed by it alone. As cells grow in masses the phenomenon of division of labor appears. The cells are arranged in separate structures or organs for special services—muscles for pulling, nerves for conducting impulses, glands for secreting. Of course, these organs are not always active. For long periods, even in the waking state, many muscles and their controlling nerves may be idle. The digestive organs do not work steadily except when given work to do. Only the respiratory organs and the heart must keep persistently at their tasks; and the heart, when beating at the moderate rate of seventy pulses per minute, is actually contracting only nine hours in the twenty-four—rest periods after each contraction amount to fifteen hours every day. Even in the parts of an organ activity is not continuous; muscle fibers take turns in keeping up a long pull, capillaries are closed down when blood is not needed, and the glomeruli of the kidneys operate in shifts. The labor of internal organs (the viscera) is, as a rule, so well regulated by inherent automatisms that the phenomena of fatigue rarely appear—the waves course over the stomach at their routine rate, the kneading movements of the intestines cannot be made to go faster than their wont. The central nervous system alone can force activities to such a degree as to bring about the limitations and inefficiencies resulting from fatigue—and that system is almost wholly limited to a control of the muscles which pull on the bones. Futhermore, fatigue itself is a check on excessive activity. It is clear, then, that the processes going on in individual cells, as well as those going on in organs, are accompanied by a large amount of local self-regulation.

III

The centrally important fact is that with the division of labor, which is implicit in the massing of cells in great multitudes and their arrangement in specific organs, most of the individual units become fixed in place so that they cannot forage for themselves. Far removed from the sources of essential supplies, these segregated and specialized units would necessarily cease their activities and would soon die unless there was developed, at the same time with their development, a means of transportation and distribution which would assure these supplies. This transporting and distributing system we recognize as the fluid matrix of the organism—the rapidly flowing blood and the more slowly moving lymph. The existence of the fluid matrix at once simplifies the problem of the remotely situated cells engaged in particular tasks. Having that provision they need not be concerned with getting food, water and oxygen, avoiding too great heat or cold, and keeping clear of the dangers of accumulating waste. All these conditions are attended to by a special organization, which, as we have seen, holds the fluid matrix constant. So long as that constancy is preserved, the various kinds of cells in the different organs are free to give full time to their special services.

The fluid matrix, therefore, is a prime requisite for the more complex organization of living units. It makes such organization possible. It gives such organization stability. And insofar as the constancy of the fluid matrix is evenly controlled it is not only a means of liberating the organism as a whole from both internal and external limitations, as we have repeatedly noted, but it is an important measure of economy, greatly minimizing the need for separate governing agencies in the various organs.

We may remark in passing that the cells in the organs which control the constancy of the fluid matrix are themselves part of the total organization of the body. They do not act by imposing conditions from the outside. In maintaining steady states in the blood and lymph they work both for the welfare of the cells in other organs essential to the body, and also for their own welfare. In short they well illustrate the arrangements for mutual dependence; in spite of generous provision for factors of safety, the integrity of the organism as a whole rests on the integrity of its individual elements, and the elements, in turn, are impotent and useless save as parts of the organized whole.

IV

In primitive conditions, small human groups living by the chase and by simple agriculture encountered circumstances not unlike those which prevail in the life of isolated single cells. Individuals were, indeed, free to move about over wide ranges and to forage for themselves, but they were dependent on what their immediate environment at the moment could furnish. They had little control over that environment. Of necessity they had to submit to the conditions which it determined.

Only when human beings are grouped in large aggregations, much as cells are grouped to form organisms, is there the opportunity of developing an internal organization which can offer mutual aid and the advantage, to many, of special individual ingenuity and skill. But with the development of larger and more complex social communities, just as with the evolution of the larger and more complex organisms, the phenomenon of division of labor becomes more and more pronounced. The list of special types of workers in a civilized society is almost unlimited. Again, like the division of labor in the animal organism, the division of labor in a complex social group has two noteworthy effects—it leads gradually to relative fixation of the individual members of the group in places where they perform their peculiar labor, and they may then be far removed from the sources of supply necessary for their continued existence. The expert mechanic in a large urban industry, for example, can neither grow his food, make his clothing, nor procure his fuel directly. He must rely on members of other groups for these things. He can do his part only so long as the others do theirs. Each one finds security in the general coöperation. Once more, just as in the body physiologic, so in the body politic, the whole and its parts are mutually dependent; the welfare of the large community and the welfare of its individual members are reciprocal.

It is obvious that at present nations have not yet achieved a full measure of success in maintaining constancy of the routine of existence or in assuring to the human elements a continuous provision for their essential needs. There is widespread search for the conditions which would diminish the anxieties and distress which are caused by the great ups and downs of economic fluctuations. Stability would free mankind from a vast amount of pain. In our own individual bodily organization we have an example of methods of successful achievement. By storage and release of material supplies, by altering the rate of continuous processes, by natural defenses against injury, and by a wide margin of safety in its functional arrangements, the normal organism protects itself for decades against perturbations. Through myriads of eons of experience our bodies, though composed of extraordinarily labile material, have developed these devices for maintaining stability. What have they to suggest?

V

At the outset it is noteworthy that the body politic itself exhibits some indications of crude automatic stabilizing processes. In the previous chapter I expressed the postulate that a certain degree of constancy in a complex system is itself evidence that agencies are acting or are ready to act to maintain that constancy. And moreover, that when a system remains steady it does so because any tendency towards change is met by increased effectiveness of the factor or factors which resist the change. Many familiar facts prove that these statements are to some degree true for society even in its present unstabilized condition. A display of conservatism excites a radical revolt and that in turn is followed by a return to conservatism. Loose government and its consequences bring the reformers into power, but their tight reins soon provoke restiveness and the desire for release. The noble enthusiasms and sacrifices of war are succeeded by moral apathy and orgies of self-indulgence. Hardly any strong tendency in a nation continues to the stage of disaster; before that extreme is reached corrective forces arise which check the tendency and they commonly prevail to such an excessive degree as themselves to cause a reaction. A study of the nature of these social swings and their reversal might lead to valuable understanding and possibly to means of more narrowly limiting the disturbances. At this point, however, we merely note that the disturbances are roughly limited, and that this limitation suggests, perhaps, the early stages of social homeostasis.

As an analogous condition of affairs, we may recall that in the evolution of vertebrate animals, and also in the development of the individual organism, the physiological devices which preserve homeostasis are at first not well developed. Only among forms which show other signs of being highly evolved do we find the automatic processes of stabilization working promptly and effectively. I would point again to the strikingly greater control of the internal environment by the complex mammalian than by the relatively simple amphibian creatures; and associated therewith the much greater freedom and independence in the presence of disturbing conditions. Is it not possible that social organization, like that of the lower animals, is still in a rudimentary stage of development? It would appear that civilized society has some of the requirements for achieving homeostasis, but that it lacks others, and because lacking them it suffers from serious and avoidable afflictions.

For the present adhering fairly strictly to physiological considerations (i.e., to the supplies of food, shelter, etc.) we are forced to recognize that the homeostasis of the individual human being is largely dependent on social homeostasis. There are certain essential needs which must be satisfied in order to preserve our personal health and efficiency. Some of the needs are satisfied gratuitously. Oxygen, and sometimes water also, we may have at will, without cost. It is noteworthy that in cities a supply of water is obtained only by community action and at public expense. There are other needs, however, which in the long run are quite as urgent as the needs for water and oxygen, and which at times cannot be satisfied because of the lack of social stability. These are the elementary requirements of food and of shelter (clothing, housing and warmth), and the benefits of medical care. To specialized workers in the social organization, limited and segregated as they are by their specialization, so that they must rely almost wholly on social homeostasis, disturbances of that homeostasis may be seriously harmful. Not only may the bodily needs be inadequately supplied, but in addition there may be suffering because of a loss of the sense of security. In the animal organism, as we have learned, the device which preserves homeostasis, which protects the cells in all parts from perturbations whether from within or without, is the controlled fluid matrix. What is the agency in civilized society which corresponds to that feature of our bodily arrangements?

VI

In a functional sense the nearest equivalent to the fluid matrix of animal organisms that is found in a state or a nation is the system of distribution in all its aspects—canals, rivers, roads and railroads, with boats, trucks and trains, serving, like the blood and lymph, as common carriers; and wholesale and retail purveyors, representing the less mobile portions of the system. In this vast and intricate stream, whose main channels and side branches reach more or less directly all communities, goods are placed, at their source, for carriage to other localities. These other localities are also sources of goods which likewise are placed in the stream. Thus the products of farm and factory, of mine and forest, are borne to and fro. But it is permissible to take goods out of the stream only if goods of equivalent value are put back in it. Ordinarily, of course, this immediate exchange does not occur. It would be highly awkward. To facilitate the process of exchange, money, which has generally recognized value, is employed. Or credit may temporarily be its substitute. By means of his money or his credit any individual can take from the stream whatever he needs or desires. Money and credit, therefore, become integral parts of the fluid matrix of society.

To assure the same degree of stability in the social organism that has been attained in the animal organism the latter suggests such control of the fluid matrix that its constancy would be maintained. That would involve, in the first instance, the certainty of continuous delivery by the moving stream of the necessities of existence. Food, clothing, shelter, the means of warmth, and assistance in case of injury or disease are naturally among these necessities. Stability would involve also the assurance of continuous remuneration of individual labor—labor which would produce exchangeable goods and which would be paid a wage sufficient to allow the laborer to take from the stream the necessary things which he and those dependent on him require. I have stated the situation, for the present, in the lowest terms. *At least* these conditions should be met if stabilization of the social organism is to be achieved. In the light of biological experience social stabilization should be sought, not in a fixed and rigid social system, but in such adaptable industrial and commercial functions as assure continuous supplies of elementary human needs.

The social organism like the animal organism is subject to disturbances, some imposed from without, some due to its own activities. Droughts, floods, earthquakes, fires and pestilence may destroy immense accumulations of goods—crops and cattle, homes and workshops—and leave great numbers of men, women and children not only destitute of the prime requirements of life, but without the means of getting them, either directly or by going to the common stream. A new machine may be invented which, because it can do the work of thousands of laborers, throws thousands of laborers out of their jobs. Thus they lose for a time the opportunity to earn the money which they must have in order to take from the stream what they require. Or there may be excessive production of certain goods so that they do not move in the stream but accumulate; or such goods have a value so much reduced that they bring little in exchange, and consequently other exchangeable goods accumulate; or men may become apprehensive of future security so that money is not used to take goods from the stream but is hoarded, and again goods accumulate; or credit may be withdrawn, which has the same effect of retarding the usual processes of trading. In whatever way the movement of goods may be checked or hindered, the result is the same. The common stream becomes clogged, its rate of flow becomes slower, manufacture becomes hazardous, workmen are therefore unemployed, and being unemployed they cannot earn the wherewithal to secure what they must have. In these various types of disaster the individual members of the social organization are not responsible for the ills which circumstance forces them to endure. As more or less fixed units, performing specialized tasks in a complex system of tasks, they are incapable of making quick adjustments to new conditions as they arise. In the emergency they are impotent to modify the system to the advantage of all. Either type of remedy—new individual adjustments or modification of the general system—requires time and thoughtful planning.

VII

What does the stability of the organism suggest as to modes of solving the problem? Here we must be careful not to extend the principles of homeostatic orderliness at first to large and unwieldy administrative regions. If we assume a limited and fairly self-sufficient administrative region, we may suppose that the suggestions of the organism would be somewhat as follows.

The organism suggests that *stability is of prime importance.* It is more important than economy. The organism throws away not only water and salts, but also sugar, if they are present in excess in the fluid matrix. This rejection is uneconomical. The organism is driven into convulsions if the sugar supply runs too low, and the convulsions mark the acme of the manoeuvres which bring forth extra sugar from the hepatic reserves to restore the normal glycemic percentage. Violent shivering may be induced to develop the additional heat which prevents a fall of body temperature. All these extreme activities, which are wasteful of energy, are not ordinarily employed, because milder measures suffice; but they are ready, whenever they are needed, to keep uniform the internal environment. This evidence that in critical times economy is secondary to stability is supported by the generous provisions of factors of safety in the body. The status of blood volume, lung capacity, blood pressure and cardiac power, for example, is not set by economy, but by the chance of having to meet unusual demands which would disturb the fluid matrix of the body if they were not met.

The organism suggests, also, that there are early signs of disturbance of homeostasis which, if sought, can be found. These warning signals are little known in the social organism, and yet their discovery and the demonstration of their real value would make contributions to social science of first-rate importance. In the complexity of modern social interrelations the strategic control would appear to reside in the devices for distributing goods, in commerce and the flow of money rather than in manufacture and production. Our bodily devices would indicate that the early warning signals, pointing to social and economic danger, should perhaps be sought in sensitive indicators of fluctuations of the commercial stream, though the causes of these fluctuations may be found in industry.

The organism suggests, furthermore, that the importance of stability warrants a specially organized control, invested by society itself with power to preserve the constancy of the fluid matrix, i.e., the processes of commerce. Does not this imply that when there is prospect of social perturbation there should be power to limit the production of goods to a degree which would reasonably adjust the supply to the demand? power to lay aside stores of goods which could be released if crises arise? power to require the accumulation of wage reserves which could be used at times of temporary unemployment? power to arrange emergency employment or training for new types of labor skill? and power to accelerate or retard the routine processes of both the production of goods and their distribution, in accordance with desirable adaptations to internal or external disturbing factors? It is noteworthy that in the bodily organism such powers as storing or releasing material reserves, hastening or checking continuous processes are exercised not by the cerebral cortex, where adaptive intelligence is mediated, but by lower centers of the brain which work in an automatic manner when appropriate signals call upon them to act.

The development of organisms indicates that the automatic devices which keep steady the internal environment have resulted from a long course of experience, possibly of experimental trial, error and correction. It seems reasonable to expect that the modes of assuring social stability, that may develop, will be the resultant of a similar evolution. Intelligence, and the example of successful stabilizing processes already in action, however, may make the evolution in society relatively rapid.

If cells of the bodily organism are injured, or are attacked by disease germs, the fluid matrix at once sets up procedures which are favorable to the restoration of the normal state. The conditions in the organism, therefore, point to the assurance of expert protective and restorative attention, through arrangements in the social group, so that the group shall not be weakened by the incapacity or ill health of its members.

We must take account of the fact that the adult organism represents a fairly fixed number of constituent cells, i.e., it is the equivalent of an adjusted population. It has no provision for any process which would be the equivalent of immigration into the social community. Nor has it any provision for unlimited growth, either as a whole or in its parts. Indeed, when some cells reproduce themselves in an uncontrolled manner they form a malignant disease, endangering the welfare of the organism as a whole. Against such pathology the body has no protection. It appears, therefore, that any wisdom which the human organism has to offer to the social organism would be based on the proviso of a population which is adjusted to reasonably assured means of subsistence and which is undisturbed by large increases from either local or foreign sources.

A noteworthy difference between the social and the biological organism is the certainty of death in the latter. In the course of existence the cells lay down intracellular substance which becomes obstructive, or they become injured by accident in irreparable ways, or they degenerate with age, until finally an essential organ of which they are members fails to play its rôle and the failure of that organ ends the activities of the whole organism. Death is a means of ridding society of old members in order to yield places for the new. A state or a nation, therefore, does not need to contemplate its own end, because its units are ceaselessly refreshed. The stabilizing processes in a body politic, therefore, when once discovered and established, might be expected to continue in operation as long as the social organization itself, to which they apply, remains fairly stable in its growth.

VIII

It is of considerable significance that the sufferings of human creatures because of lack of stability in the social organism have more and more stimulated efforts directed towards improvement. Various schemes for the avoidance of economic calamities have been put forth not only by dreamers of Utopias but also by sociologists, economists, statesmen, labor leaders and experienced managers of affairs. In all such proposals a much greater control of credit, currency, production, distribution, wages and workmen's welfare is anticipated than has been regarded as expedient or justifiable in the individualistic enterprises of the past.

Communists have offered their solution of the problem and are trying out their ideas on a large scale in Soviet Russia. The socialists have other plans for the mitigation of the economic ills of mankind. And in the United States, where neither communism nor socialism has been influential, various suggestions have been offered for stabilizing the conditions of industry and commerce. Among these suggestions are the establishment of a national economic council or a business congress or a board of industries or of trade associations, representing key industries or the more highly concentrated industries, and endowed (in some of the schemes) with mandatory power to coördinate production and consumption for the benefit of wage earners; provision for regularity and continuity of employment, with national employment bureaus as an aid, with unemployment insurance as a safety device, and with planned public works as a means of absorbing idle workmen; incentives for the preservation of individual initiative and originality in spite of the dangers of fixed organization; shortening of the working time and prohibition of child labor; the raising of the average industrial wage; and the assuring of the general public through governmental regulation that in any arrangements which are made its interests will be protected.

The multiplicity of these schemes is itself proof that no satisfactory single scheme has been suggested by anybody. The projection of the schemes, however, is clear evidence that in the minds of thoughtful and responsible men a belief exists that intelligence applied to social instability can lessen the hardships which result from technological advances, unlimited competition and the relatively free play of selfish interests.

By application of intelligence to medico-social problems, destructive epidemics such as the plague and smallpox have been abolished; fatal afflictions, e.g., diphtheria and tuberculosis, have been greatly mitigated and largely reduced; and vast areas of the earth's surface, formerly dangerous to man, have been made fit for safe and sanitary habitation because of the conquest of malaria, yellow fever and hookworm disease. These achievements all involve social organization, social control, and a lessening of the independence of the individual members. Economic and sociological programs, in which emphasis is laid on the well-being of the human elements in production as well as on material profits, have purposes similar to the medical programs just mentioned. They recognize that the social organism, like the bodily organism, cannot be vigorous and efficient unless its elements are assured the essential minimal conditions for healthful life and activity. And the possession of a mind by the human elements would require that these conditions include not only provision for the elementary needs which we have been considering, but also reasonable satisfaction of desires.

IX

In our study of the effects on the organism of a controlled stability of the fluid matrix we noted that just insofar as the stability is preserved the organism is released from the limitations imposed by internal and external disturbances. Is it not probable that similar results will flow from control and stabilization of the fluid matrix of the social organism? The hope is not unreasonable that the distress arising from catastrophes can be greatly mitigated, and that the suffering due to lack of necessary things which is attendant on great economic fluctations can be obviated, by carefully planning and by intelligently regulating the processes of production and distribution. Banishment of this distress and suffering would bring freedom from fears, worries and anxieties concerning livelihood, which now may fill men with dark despair. As a Lord Chancellor of England has declared, and his declaration has been approved by a Justice of the United States Supreme Court, "Necessitous men are not, truly speaking, free men." The assurance of freedom *to men who are willing to work* would justify a larger control of economic processes, repugnant though that may seem, for it would be a sacrifice of lesser for greater values.

Bodily homeostasis, as we have learned, results in liberating those functions of the nervous system that adapt the organism to new situations, from the necessity of paying routine attention to the management of the details of bare existence. Without homeostatic devices we should be in constant danger of disaster, unless we were always on the alert to correct voluntarily what normally is corrected automatically. With homeostatic devices, however, that keep essential bodily processes steady, we as individuals are free from such slavery—free to enter into agreeable relations with our fellows, free to enjoy beautiful things, to explore and understand the wonders of the world about us, to develop new ideas and interests, and to work and play, untrammeled by anxieties concerning our bodily affairs. The main service of social homeostasis would be to support bodily homeostasis. It would therefore help to release the highest activities of the nervous system for adventure and achievement. With essential needs assured, the priceless unessentials could be freely sought.

There might be apprehension that social stabilization

would tend towards dull monotony, that the excitements of uncertainty would be lacking. That would be true, however, only for the fundamental requirements of existence. There would still be the social disturbances of new inventions, the social interest in renowned exploits, in the discords of human nature, in reports of fresh ideas, in the intrigues of love and hate, and in whatever other events there may be that make life varied and colorful. Above all there might be apprehension that social stabilization would too greatly interfere with the free action of individuals. As repeatedly emphasized, however, steady states in society as a whole and steady states in its members are closely linked. Just as social stabilization would foster the stability, both physical and mental, of the members of the social organism, so likewise it would foster their higher freedom, giving them serenity and leisure, which are the primary conditions for wholesome recreation, for the discovery of a satisfactory and invigorating social *milieu*, and for the discipline and enjoyment of individual aptitudes.

A LIST OF PUBLICATIONS FROM THE PHYSIOLOGICAL LABORATORY IN HARVARD UNIVERSITY, ON WHICH THE PRESENT ACCOUNT IS BASED

1. The Movements of the Stomach and Intestines in Some Surgical Conditions. By W. B. Cannon and F. T. Murphy. Annals of Surgery, 1906, xliii, pp. 513–536.
2. The Mechanical Factors of Digestion. By W. B. Cannon. London, 1911.
3. An Explanation of Hunger. By W. B. Cannon and A. L. Washburn. American Journal of Physiology, 1912, xxix, pp. 441–454.
4. The Interrelation of Emotions as Suggested by Recent Physiological Researches. By W. B. Cannon. American Journal of Psychology, 1914, xxv, pp. 256–282.
5. The Hastening of Coagulation by Stimulating the Splanchnic Nerves. By W. B. Cannon and W. L. Mendenhall. American Journal of Physiology, 1914, xxxiv, pp. 243–250.
6. The Effects of Hemorrhage Before and After Exclusion of Abdominal Circulation, Adrenals, or Intestines. By H. Gray and L. K. Lunt. *Ibid.*, 1914, xxxiv, pp. 332–351.
7. The Physiological Basis of Thirst. The Croonian Lecture of the Royal Society, London. By W. B. Cannon. Proceedings of the Royal Society, 1918, xc (B), pp. 283–301.
8. The Nature and Treatment of Wound Shock and Allied Conditions. By W. B. Cannon, J. Fraser and A. N. Hooper. Journal of the American Medical Association, 1918, lxx, pp. 607–621.
9. The Basal Metabolism in Traumatic Shock. By J. C. Aub. American Journal of Physiology, 1920, liv, pp. 388–407.
10. Further Observations on the Denervated Heart in Relation to Adrenal Secretion. By W. B. Cannon and D. Rapport. *Ibid.*, 1921, lviii, pp. 308–337.
11. The Metabolic Effect of Adrenalectomy upon the Urethanized Cat. By J. C. Aub, E. M. Bright and J. Forman. *Ibid.*, 1922, lxi, pp. 349–368.
12. The Critical Level in a Falling Blood Pressure. By W. B. Cannon and McKeen Cattell. Archives of Surgery, 1922, iv, pp. 300–323.
13. Traumatic Shock. By W. B. Cannon. D. Appleton and Company., New York, 1923.
14. The Effects of Muscle Metabolites on Adrenal Secretion. By W. B. Cannon, J. R. Linton and R. R. Linton. American Journal of Physiology, 1924, lxxi, pp. 153–162.
15. A Sympathetic and Adrenal Mechanism for Mobilizing Sugar in Hypoglycemia. By W. B. Cannon, M. A. McIver and S. W. Bliss. *Ibid.*, 1924, lxix, pp. 46–66.
16. Changes in Metabolism Following Adrenal Stimulation. By M. A. McIver and E. M. Bright. *Ibid.*, 1924, lxviii, pp. 622–644.

17. The Nervous Control of Insulin Secretion. By S. W. Britton. *Ibid.*, 1925, lxxiv, pp. 291–308.
18. The Rôle of the Adrenal Medulla in Pseudaffective Hyperglycemia. By E. Bulatao and W. B. Cannon. *Ibid.*, 1925, lxxii, pp. 295–313.
19. Pseudaffective Medulliadrenal Secretion. By W. B. Cannon and S. W. Britton. *Ibid.*, 1925, lxxii, pp. 283–294.
20. Some General Features of Endocrine Influence on Metabolism. By W. B. Cannon. American Journal of Medical Sciences, 1926, clxxi, pp. 1–20.
21. A Lasting Preparation of the Denervated Heart for Detecting Internal Secretion. By W. B. Cannon, J. T. Lewis and S. W. Britton. American Journal of Physiology, 1926, lxxvii, pp. 326–352.
22. The Influence of Motion and Emotion on Medulliadrenal Secretion. By W. B. Cannon and S. W. Britton. *Ibid.*, 1927, lxxix, pp. 433–465.
23. The Rôle of Adrenal Secretion in the Chemical Control of Body Temperature. By W. B. Cannon, A. Querido, S. W. Britton and E. M. Bright. *Ibid.*, 1927, lxxix, pp. 466–507.
24. Die Notfallsfunktionen des Sympathico-adrenalen Systems. By W. B. Cannon. Ergebnisse der Physiologie, 1928, xxvii, pp 380–406.
25. Emotional Polycythemia in Relation to Sympathetic and Medulliadrenal Action on the Spleen. By J. J. Izquierdo and W. B. Cannon. American Journal of Physiology, 1928, lxxxiv, pp. 545–562.
26. Emotional Relative Mononucleosis. By V. Menkin. *Ibid.*, 1928, lxxxv, pp. 489–497.
27. Reasons for Optimism in the Care of the Sick. By W. B. Cannon. New England Journal of Medicine, 1928, cxcix, pp. 593–597.
28. A Diencephalic Mechanism for the Expression of Rage, with Special Reference to the Sympathetic Nervous System. By P. Bard. American Journal of Physiology, 1928, lxxxiv, pp. 490–516.
29. Some Conditions Affecting the Capacity for Prolonged Muscular Work. By F. A. Campos, W. B. Cannon, H. Lundin and T. T. Walker. *Ibid.*, 1929, lxxxvii, pp. 680–701.
30. Bodily Changes in Pain, Hunger, Fear and Rage. By W. B. Cannon. D. Appleton and Company, New York, 2nd Edition, 1929.
31. Organization for Physiological Homeostasis. By W. B. Cannon. Physiological Reviews, 1929, ix, pp. 399–431.
32. Some Aspects of the Physiology of Animals Surviving Complete Exclusion of Sympathetic Nerve Impulses. By W. B. Cannon, H. F. Newton, E. M. Bright, V. Menkin and R. M. Moore. American Journal of Physiology, 1929, lxxxix, pp. 84–107.
33. The Autonomic Nervous System, an Interpretation. The Linacre Lecture. By W. B. Cannon. The Lancet, 1930, i, pp. 1109–1115.
34. Blood Calcium after Sympathectomy, Adrenin and Sham Rage. By J. Lamelas. American Journal of Physiology, 1930, xciii, pp. 111–115.
35. Observations on the Central Control of Shivering and of Heat Regulation in the Rabbit. By S. Dworkin. *Ibid.*, 1930, xciii, pp. 227–244.
36. Impotence of the Male Rodent after Sympathetic Denervation of the Genital Organs. By Z. M. Bacq. *Ibid.*, 1931, xcvi, pp. 321–330.
37. The Effects of Progressive Sympathectomy on Blood Pressure. By Bradford Cannon. *Ibid.*, 1931, xcvii, pp. 592–595.
38. A Belated Effect of Sympathectomy on Lactation. By W. B. Cannon and E. M. Bright. *Ibid.*, 1931, xcvii, pp. 319–321.
39. A Method for Uniform Stimulation of the Salivary Glands in the Un-

anesthetized Dog by Exposure to a Warm Environment, with Some Observations on the Quantitative Changes in Salivary Flow during Dehydration. By M. I. Gregersen. *Ibid.*, 1931, xcvii, pp. 107–116.

40. Reflex Stimulation and Inhibition of Vasodilators in Sympathectomized Animals. By N. E. Freeman and A. Rosenblueth. *Ibid.*, 1931, xcviii, pp. 454–462.
41. The Effect of Sympathectomy on Sexual Functions, Lactation, and the Maternal Behavior of the Albino Rat. By Z. M. Bacq. *Ibid.*, 1932, xcix, pp. 444–453.
42. Studies of Homeostasis in Normal, Sympathectomized and Ergotaminized Animals. I. The Effect of High and Low Temperatures. By M. E. MacK. Sawyer and T. Schlossberg. *Ibid.*, 1933, civ, pp. 172–183.
43. II. The Effect of Anoxemia. By M. E. MacK. Sawyer, T. Schlossberg and E. M. Bixby. *Ibid.*, 1933, civ, pp. 184–189.
44. III. The Effect of Insulin. By T. Schlossberg, M. E. MacK. Sawyer and E. M. Bixby. *Ibid.*, 1933, civ, pp. 190–194.
45. IV. The Effect of Hemorrhage. By T. Schlossberg and M. E. MacK. Sawyer. *Ibid.*, 1933, civ, pp. 195–203.
46. A Further Study of Vasodilators in Sympathectomized Animals. By A. Rosenblueth and W. B. Cannon. *Ibid.*, 1934, cviii, pp. 599–607.
47. Stresses and Strains of Homeostasis. By W. B. Cannon. Am. Journ. Med. Sci., 1935, clxxxix, pp. 1–14.
48. The Heart Rate of the Sympathectomized Dog in Rest and Exercise. By L. Brouha, W. B. Cannon and D. B. Dill. Journ. Physiol., 1936, lxxxvii, pp. 345–359.
49. The Effect of Anterior Hypophyseal Extract upon the Serum Calcium and Phosphorus. By H. B. Friedgood and R. McLean. Am. Journ. Physiol., 1937, cxviii, pp. 588–593.
50. Effect of Sympathin on Blood Sugar. By R. C. Bodo and A. E. Benaglia. *Ibid.*, 1938, cxxi, pp. 728–737.
51. The Effect of Sympathectomy on Gestation and Lactation in the Cat. By F. A. Simeone and J. F. Ross. *Ibid.*, 1938, cxxii, pp. 659–667.
52. Studies of Homeostasis in the Sympathectomized Dog. By F. K. McDonough. *Ibid.*, 1939, cxxv, pp. 530–546.
53. Blood-Sugar Variations in Normal and in Sympathectomized Dogs. By L. Brouha, W. B. Cannon and D. B. Dill. Journ. Physiol. (In press).

INDEX